D0855327

Richard H. Brown and Paul E. Cohen

Revolution

MAPPING THE ROAD TO AMERICAN INDEPENDENCE 1755–1783

W. W. NORTON & COMPANY
NEW YORK LONDON

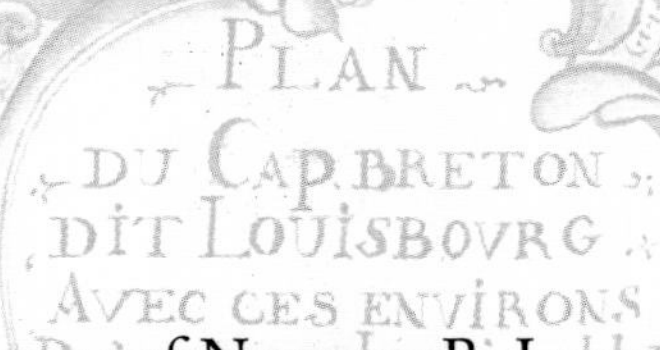

In memory of Norman B. Leventhal
(1917–2015)
who inspired us both

REVOLUTION: MAPPING THE ROAD TO AMERICAN INDEPENDENCE 1755–1783

Printed in China

FIRST EDITION

For information about special discounts for bulk purchases, please contact W. W. Norton Special Sales at specialsales@wwnorton.com or 800-233-4830

Manufacturing by South China, a division of R. R. Donnelley China

Book design and composition by Laura Lindgren

Production manager: Julia Druskin

Library of Congress Cataloging-in-Publication Data
Brown, Richard H. (Map collector), author.
Revolution : mapping the road to American independence, 1755–1783 / Richard H. Brown and Paul E. Cohen. – First edition.
pages cm
"Copyright © 2015 by Richard H. Brown and Paul E. Cohen."
Includes bibliographical references and index.
ISBN 978-0-393-06032-4 (hardcover)
1. United States–History–French and Indian War, 1754–1763–Maps. 2. United States–History–Revolution, 1775–1783–Maps. 3. United States–History–Revolution, 1775–1783–Cartography. 4. United States–History–18th century–Maps. 5. United States–History–18th century–Cartography. I. Cohen, Paul E., author. II. Title.
G1201.S26B7 2015
973.3022'3–dc23 2015009834

frontispiece: Detail, *The Death of General Wolfe*, Benjamin West, 1770, oil on canvas, 153 × 215 cm. National Gallery of Canada, Ottawa (figure 18, p. 42)

pages iv–v: Detail, *Plan du Cap Breton, dit Louisbourg avec ces environs*, cartographer unknown, 1758, manuscript, 51 × 203 cm. Geography and Map Division, Library of Congress (figure 14, pp. 30–31)

W. W. Norton & Company, 500 Fifth Avenue, New York, NY 10110

www.wwnorton.com

W. W. Norton & Company Ltd., Castle House, 75/76 Wells Street, London, WIT 3QT

3 4 5 6 7 8 9 0

Contents

Ship Yards
Brookland Ferry
Remsen's Mill
The WALLABOUT BAY
Outward Boundaries
Southward Boundaries
Westward Boundaries
Distillery
Phil. Livingston Esq.
The Governours or Nutten Iſland
Red Hook
Mill Dam
BROOKLAND Parish
Road to Jamaica
Road to Flatbush
BEDFORD
PART OF
Scale of 5000 Feet.
Scale of One Mile.
Scale of Yards.
A South West View of the City of New York. Taken from the Governours Iſland at *
Long Island

Introduction

Revolution: Mapping the Road to American Independence is the story of warfare in America as told through a series of historic maps. Our account begins twenty years before the skirmishes at Lexington and Concord in 1775, which mark the onset of the Revolutionary War. The French and Indian War was fought from 1755 to 1763, and it set the stage for the Revolution. That earlier conflict is vague in the minds of many, as few people know when or why it was fought. In fact, this war was bloodier than the Revolution, took more lives, and in many ways was more significant. When the dust settled, France had lost an entire continent, and Britain had gained clear title to most of North America.

The cost of waging this seven-year war on a distant continent had nearly bankrupted England. Increasing taxes was one solution to the country's financial woes, but when England began to impose new taxes on its colonists the seeds of revolution were planted. Disagreements between the rulers and the ruled led to the sequel war and eventually independence for the colonists. Many of the same people participated in both wars: George Washington cut his teeth in the French and Indian War, often fighting alongside fellow soldiers who would become his enemies a few years later.

Most historians seek out maps to illustrate and support their narratives. Our narrative supports the maps. As map historian Lloyd Brown has written, "Pictorial news about the war was limited almost entirely to maps and as source material on the Revolution these are of the utmost importance." Maps do more than report news of the war; they provide the fullest pictorial record of America during the second half of the eighteenth century.

In *Revolution*, we have tracked down maps, and some perspective views, that bring visual energy to the defining battles. More has been written about the Revolutionary War than the French and Indian War, and many more maps from it have been produced. Early on, we discovered that the maps of the French and Indian War were usually of fortifications and often had a similar look. We had to step up our search in order to find works that showed the actual battles of the earlier war. In the process, we located many that had never been reproduced before. These include maps in libraries but also some in private collections.

Detail, figure 22. *Plan of the City of New York in North America: Surveyed in the Years 1776 and 1777*, Bernard Ratzer, London, ca. 1770, p. 47

In our determination to locate the maps that seem closest to the actions portrayed, we discovered that the most candid maps were hand-drawn works executed by eyewitnesses. Manuscript maps seem to bring us closer to actual events, and some of these have only recently come to light. While not all of the maps in *Revolution* are manuscripts, we have focused on works in this form. In some cases they were later engraved by printers and became published maps but in most cases they were never printed.

A large group of maps in this book is from the King George III (1738–1820) collection at the British Library. Throughout his sixty-year reign, the king collected maps and views with passion and would often disappear into his study for hours on end in order to spend time with them. His son King George IV (1762–1830) bequeathed his books and maps to the British Museum (now the British Library). It is so large and valuable that Peter Barber, the Head of Map Collections at the British Library, has stated that if that national institution consisted exclusively of George III's collection, it would still be a world-class library. The collection is so enormous that much of it has only recently been cataloged and only a fraction of the maps have ever been reproduced. In *Revolution* we have twelve maps from the British Library.

A remarkable collection of Revolutionary War maps can also be found at Alnwick Castle in England, and we traveled to Northumberland to see the Lord Percy collection. Percy was a general in the early battles of the war and a man with a keen interest in cartography. He even employed Claude Joseph Sauthier, one of the leading cartographers of the era, as his secretary. The collection at Alnwick is undoubtedly the finest in private hands, and it is represented in *Revolution* by eight maps.

The Library of Congress was also a rich source, and there are more maps in *Revolution* from that institution than from any other. William Faden was the principal printer of maps during the Revolutionary War and the library possesses his archive. It consists of manuscripts that he used to prepare maps for the press, along with many printed maps. In addition, the library has the Rochambeau collection. General Rochambeau was part of the French alliance that defeated the British at Yorktown. Many manuscripts of the war in the book come from the Faden and Rochambeau collections.

Some key maps remained in the hands of the participants themselves. British general William Clinton's maps and papers were retained by his family and were acquired by the Clements Library at the University of Michigan. General Lafayette's maps also remained in his family and these trickled onto the market during the twentieth century. And some lesser figures also had significant maps that have recently turned up. The most important map of the New York campaign in 1776 and another map of Philadelphia had been in the possession of a lesser-known British general, Sir William Erskine. When they appeared at auction in 2010, no one had previously known of their existence. Both are in *Revolution*.

The campaigns of these two wars took place in the interior of North America and up and down the Atlantic coast from Nova Scotia to the Caribbean. When the French and Indian War ended, the entire east coast of America was British and the crown jewel of the British Empire had been secured. When the Revolution ended two decades later, Britain had lost its prize possession. Enormous change took place in North America between 1755 and 1783, and the events of these years were eagerly watched around the world. For that reason there is a cosmopolitan array of maps in this book–from Germany, Canada, France, England, and America. By gathering works from these places we have added an international dimension to this most decisive period in American history. Within thirty years, the French and Indian and Revolutionary wars were fought and the United States had become a nation.

Detail, figure 71. *A Map of the British Colonies in North America*, John Mitchell, London, 1755/1775, p. 139

pages 4–5: Detail, figure 13. *Carte Topographique du Port et la Ville de Louisbourg*, Pierre-Jerome Lartigue, 1758, pp. 28–29

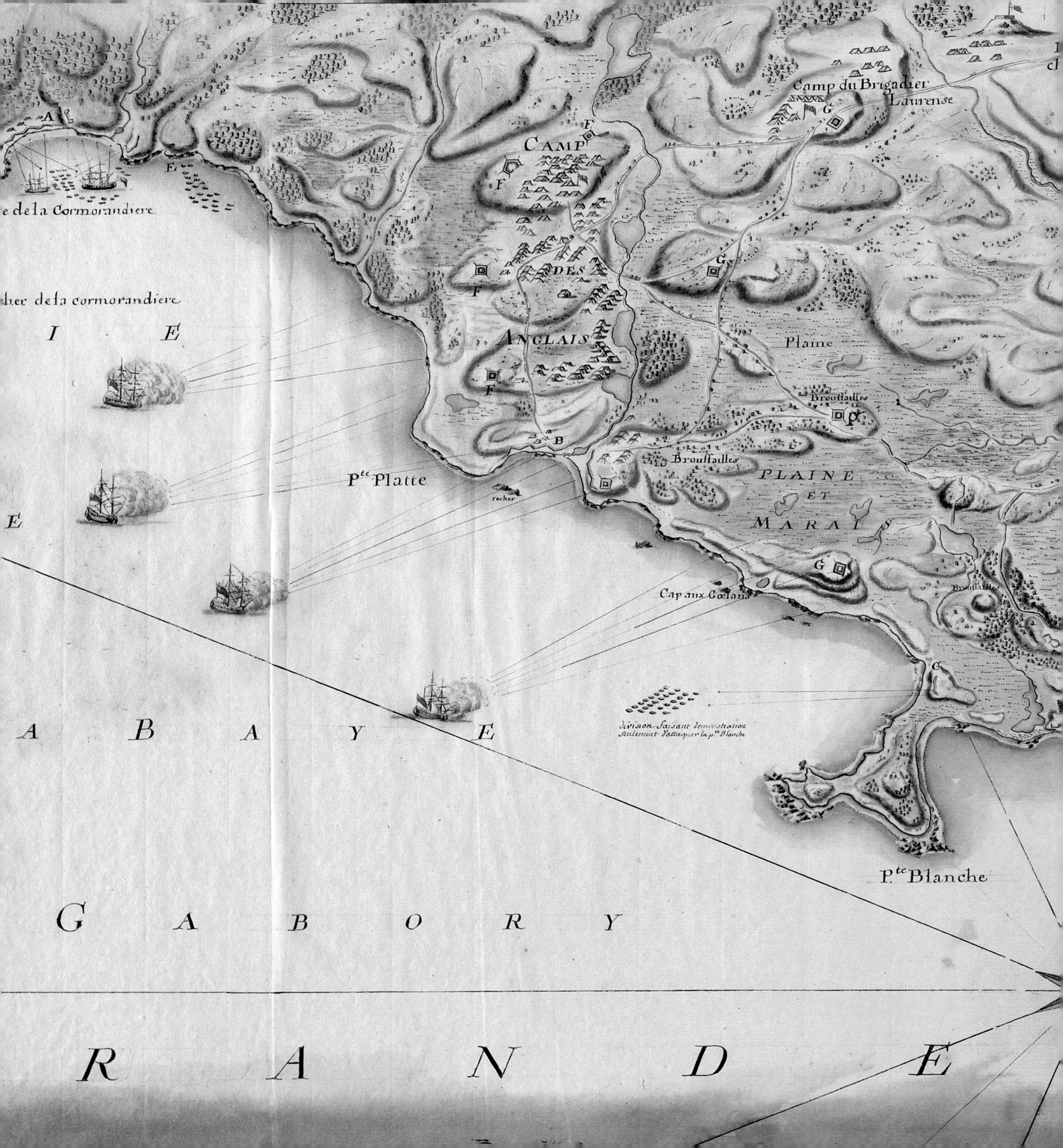
de la Cormorandiere
cher de la cormorandiere
Camp du Brigadier Laurense
CAMP DES ANGLAIS
Plaine
Brouffailles
Brouffailles
PLAINE ET MARAIS
Pte Platte
rocher
Cap aux Gœlans
Brouffailles
Division faisant demonstration seulement d'attaquer la Pte Blanche
Pte Blanche
A B A Y E
G A B O R Y
R A N D E

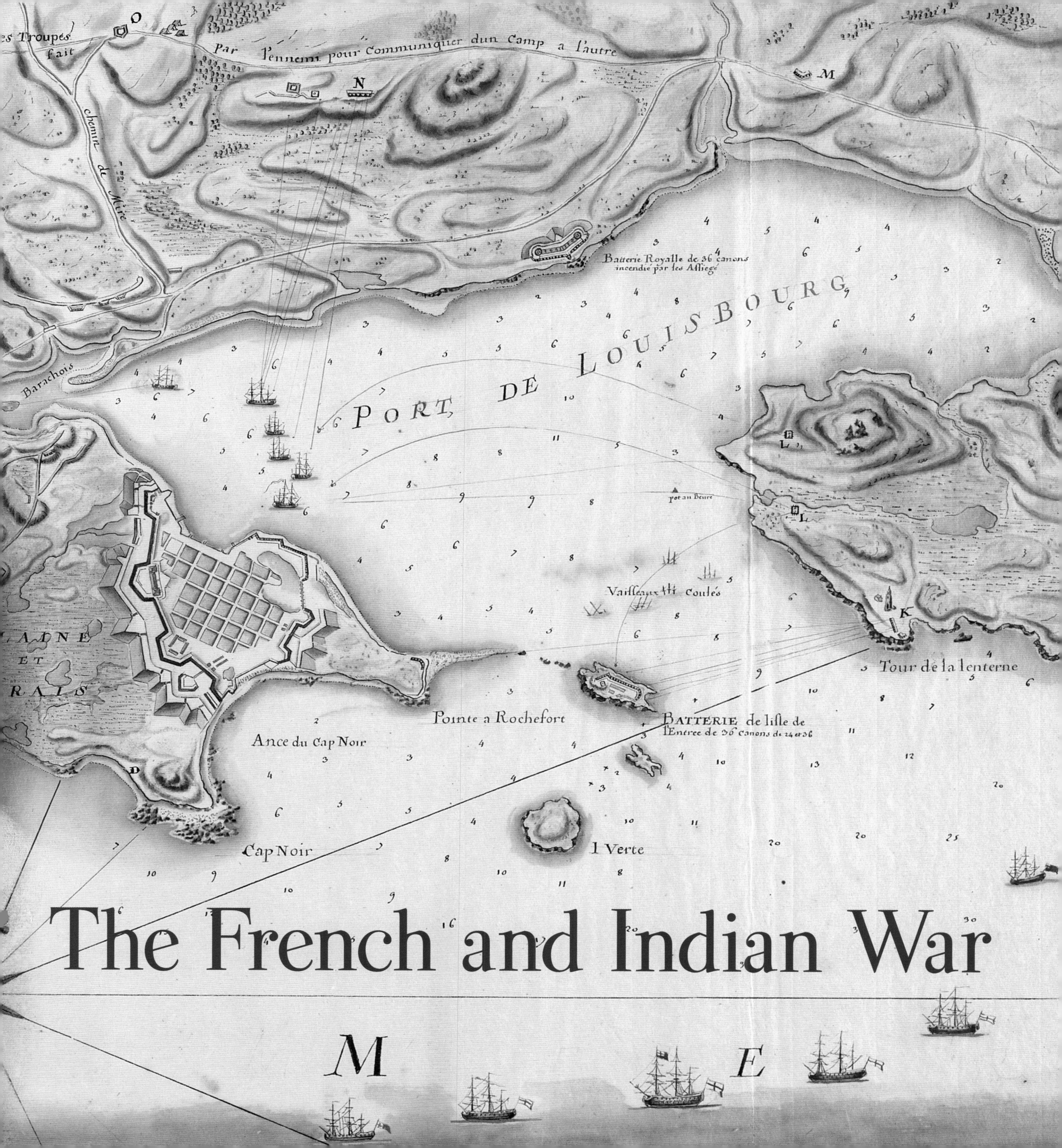

The French and Indian War

THE ENGLISH EMPIRE IN NORTH AMERICA:

ters, *and the* formal Surrender *of their* INDIAN FRIENDS;

CH, *with the several* Forts *they have* unjustly erected *therein*. By a Society of Anti-Gallica

Longitude West from 35 Ferro

Longitude West from London

LABRADOR or

Little Esquimaux

NEW BRITAIN

Treaty of Utrecht

L. Mistassin

Ft. Rupert

Rupert R.

Lake Peritibi

Manikouagane

R. Bustard

Lake Natkatich

French Factory

Tiekouagami

Saguenay R.

L. St. Jean

Tadoussac

L. Begon

L. Oumaniouetz

CANADA

Lake Temiscaming

Trois Rivieres

QUEBEC

L. St. Pierre

R. Outaouais

Montreal

Iroquois

L. St. Francis

Riv. St. Laurence

Sept Isles

Anticosti

Gaspesia

C. Rosiers

Gaspe B.

I. Bonaventura

Bay des Chaleurs

GULF OF St. LAURENCE

B. St. George

Caraquet I.

Miscou I.

Miramichi B.

Madelaine Isles

C. Ray

St. Paul

C. BRETON I.

NOVA SCOTIA

Micmacs

Souriquois

Chignecto

Milford

Canso

Fundy B.

Annapolis Royal

Lunenburg

Halifax

Chebucto B.

C. Sable

Browns Bank

I. of Sable

I. of Sable Bank

St. Peter's Bank

Middle Bank

Fort Bank

Canso Bank

Bank

New found

St. John

Sagadahook

TERRITORY

Abnakis

NEW ENGLAND

NEW HAMPSHIRE

MASSACHUSETS

CONNECTICUT

Boston Harb.

BOSTON

C. Cod

Chatham

Sandy Pt.

Nantucket I.

Martha's Vineyard I.

Providence

Newbury

Salem

Salisbury

R. of Piscataqua

York

L. de Champlain

Ft. Frederic Crown Point

NEW YORK

Albany

Schenectady

Oswego

Oneida

Cayugas

Tuscarora

LAKE ONTARIO

LONG I.

N. YORK

Amboy

Shrewsbury

PHILADELPHIA

Burlington

JERSEY

Barnegat Beach

Absecon Beach

Ludleys Beach

C. May

Delaware R.

C. Hinlopen

Cedar I.

Swansecut Gr.

Taches I.

C. Charles

MARYLAND

SYLVANIA

ATLANTIC OCEAN

NEW FOUNDLAND

C. Bonavista

Trinity Bay

Conception Bay

St. John

C. Race

Placentia

B. of Placentia

Port H.

The 3 Isles

Green B.

White B.

Pingouin I.

Belle Isle

Quirpon

St. Julian

Bell Isle

Goboso Bay

C. Richi

B. St. Paul

Ferrol Pt.

C. Charles

R. St. Francis

The several Provinces of
English Empire in N. Am
are distingushed by Red, B
Green and Yallow accordin
their respective Jurisdictio
The Spanish by Brown;
the French Possessions and
-croachments are without
Colour.

HUDSONS BAY

F. Pr. of Wales

C. Churchill

C. Totnam

Churchill

York F.

P. Nelson R.

New Severn

C. Henr. Maria

NEW SOUTH WALES

JAMES'S BAY

C. Jones

Albany R.

Moose R.

Ft. Rupert

L. Mistassin

C. Smith

Non Discover'd Sea

C. Chidley

LABRADOR or NEW BRITAIN

Slude R.

B. Eskimaux

Anticosti

The Anti-Gallican Map

Cartography always benefits from war or even the prospect of war. Sometimes maps are even the cause of warfare. During the middle years of the eighteenth century, tensions over dominance in North America flared as British and French cartographers each claimed large, overlapping territories for their respective colonies on the continent. The frenzy reached its peak in 1755 with the publication of John Mitchell's *Map of the British and French Dominions in North America,* which would serve as the template map of America for the next thirty years. But another map, little recognized by scholars or collectors, had no match in its effectiveness to draw Britain and France into war. *A New and Accurate Map of the English Empire in North America . . . by a Society of Anti-Gallicans* (figure 1) incorporated Mitchell's geographic improvements but extended and dramatized his territorial claims. The next year, the two most powerful countries on earth were fighting a war that many historians consider among the most decisive in all of history.

Robert Sayer and William Herbert published the "Anti-Gallican" map in December 1755, and it was a careful synthesis of already existing publications. The impetus was the activity of the Society of Anti-Gallicans. The result was a masterpiece of propaganda. As Mary Pedley of the Clements Library observed: "By translating a complicated and unresolved political and military situation into the black, white and colored graphics of the map, the Anti-Gallican designer has played on the fears and prejudices of the public to encourage the necessity of war."

The Laudable Society of Anti-Gallicans was founded in 1745 and attracted "Gentlemen of the Best characters and address," including established merchants, tradesmen, and military luminaries, to its frequent meetings throughout London. Its mission was "to promote the British Manufactures, to extend the Commerce of *England*, and discourage the introducing of *French Modes,* and oppose the importation of *French Commodities.*" Members or "Brethren" wore enameled badges of St. George spearing the French flag from the saddle of his horse. The badges of the society's Grand Presidents were particularly ornate and one example in the British Museum (figure 2) is labeled "remarkable as a rare survival of a piece of jewelry in the Rococco style." Badges were fabricated at the Battersea Enamel Works and their scarcity may be attributed to the short life of that

Opposite (detail) and pages 8–9: Figure 1. *A New and Accurate Map of the English Empire in North America*, Society of Anti-Gallicans, London, 1755, copperplate engraving, 41 × 50 cm. Library and Archives Canada/Maps, Plans and Charts collection/NMC21053.gvv

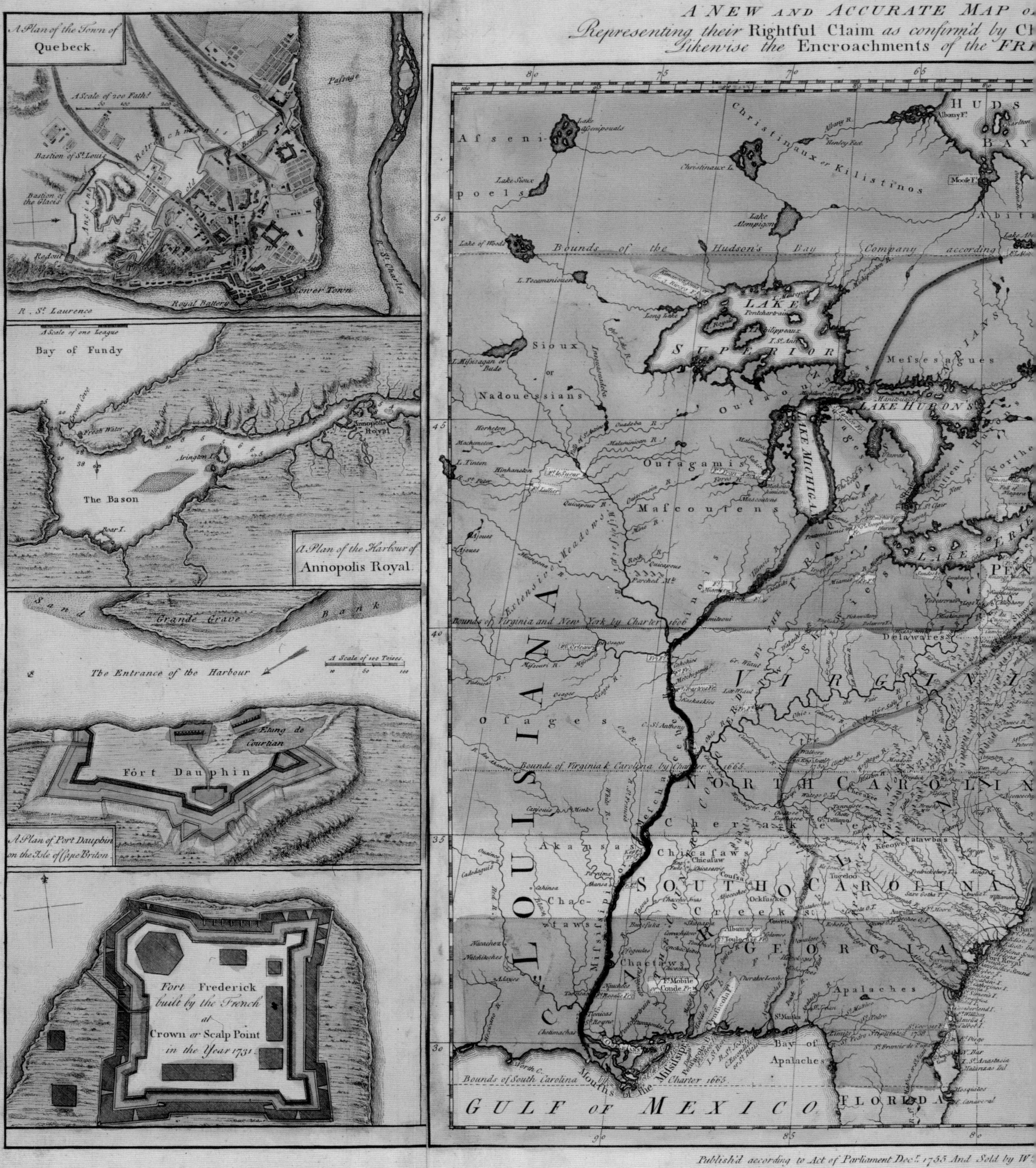

A NEW AND ACCURATE MAP o
Representing their Rightful Claim as confirm'd by Ch
Likewise the Encroachments of the FR
A Plan of the Town of Quebeck.
A Scale of 200 Fath.s
Bastion of St. Louis
Bastion of the Glacis
Retrenchments
Passage
Lower Town
Royal Battery
R. St. Laurence
R. St. Charles
A Scale of one League
Bay of Fundy
Annopolis Royal
Aringen I.
The Bason
Boar I.
A Plan of the Harbour of Annopolis Royal.
Sand Bank
Grande Grave
The Entrance of the Harbour
A Scale of 100 Toises
Etang de Courtian
Fort Dauphin
A Plan of Port Dauphin on the Isle of Cape Briton
Fort Frederick built by the French at Crown or Scalp Point in the Year 1731.
HUDS BAY
Christinaux or Kilistinos
Lake Alempigon
Bounds of the Hudson's Bay Company according
LAKE SUPERIOR
Sioux or Nadouessians
Outagamis
Mascoutens
LAKE MICHIGAN
LAKE HURON'S
LAKE ERIE
Bounds of Virginia and New York by Charter 1606
LOUISIANA
Osages
VIRGINIA
Bounds of Virginia & Carolina by Charter 1665
NORTH CAROLINA
Cherakee
Chicafaws
SOUTH CAROLINA
Creeks
GEORGIA
Chactaws
Apalaches
Bay of Apalaches
Mouths of the Missisipi
Bounds of South Carolina
Charter 1665
GULF OF MEXICO
FLORIDA
Publish'd according to Act of Parliament Dec.r 1755 And Sold by W.

ENGLISH EMPIRE IN NORTH AMERICA:
and the formal Surrender of their INDIAN FRIENDS;
with the several Forts they have unjustly erected therein. By a Society of Anti-Gallicans.
Longitude West from 35 Ferro
Longitude West from London
LABRADOR or
Little Esquimaux
NEW BRITAIN
NEW FOUND LAND
GULF OF ST. LAURENCE
C. BRETON I.
Great Bank of Newfoundland
OCEAN
ATLANTIC
THE WESTERN OR ATLANTIC
MASSACHUSETS
NEW YORK
C. Cod
C. Hatteras
The several Provinces of the English Empire in N. America are distinguished by Red, Blue, Green and Yellow according to their respective Jurisdictions. The Spanish by Brown; but the French Possessions and Encroachments are without any Colour.
The French claim all the Country within the Hudson's Bay Company's Southern Limits and the Brown Line.
The Purple Line represents the Western Boundary of the hereditary & Conquer'd Country of our Indian Friends & Allies, which has been ceded and confirm'd to us by several Treaties and Deeds of Sale.
A Plan of the Harbour and Town of Louisbourg on the Isle of Cape Briton.
A Scale of 500 Toises.
A Plan of CHEBUCTO HARBOUR
Scale of Miles
BEDFORD BAY
Chebucto Harbour
HUDSONS BAY
LABRADOR or NEW BRITAIN
NEW SOUTH WALES
CANADA
LOUISIANA
VIRGINIA
CAROLINA
GEORGIA
CAROLANA
GULF OF MEXICO
Havanah
MEXICO or NEW SPAIN
Bay of Honduras
Jamaica
PACIFIC OCEAN OR SOUTH SEA
Bermudas
THE ATLANTIC OCEAN
C. Farewell
Azores or Western Isles
I. Madeira
Canary Is.
C. Verd
IRELAND
SCOTLAND
ENGLAND
FRANCE
SPAIN
PORTUGAL
Bay of Biscay
FEZ
MAROKKO
AFRICA
London Bridge & Robt. Sayer over against Fetter Lane in Fleet Street.

facility, which was founded in 1753 by Stephen Janssen, the society's president and the Lord Mayor of London, but failed in 1756 in spite of its revolutionary process for transferring images onto enamel.

Figure 2. Badge of a Grand President of the Anti-Gallican Society, ca. 1750. © Trustees of the British Museum

Initially the society's activities amounted to little more than boycotting claret and French lace, but international relations became increasingly unsettled in the years following the 1748 Treaty of Aix-la-Chapelle, which ended the War of the Austrian Succession but not the territorial disputes in America. The society sponsored prizes for English products that could stand in for their French counterparts such as lace ruffles and for commercial achievements, particularly in support of the English fishing industry. A frustrated French minister of war, Comte D'Argenson, complained that "Their [British] company of Anti-Gallicans encourages their fischeries in order to destroy ours," adding, "It is a great mark of national enmity to have given publicily such a title to that company."

Anti-Gallicans went from boycotting lace to advocating martial force. When war was declared in 1756, the society immediately outfitted a privateer, belligerently named the *Anti-Gallican*, which soon managed to engage a French vessel off the Spanish coast. After a four-hour exchange of cannon with considerable loss of life, the *Anti-Gallican* triumphantly towed its prize into the port of Cádiz. The Spanish took a dim view of this conquest, claiming the engagement had occurred in their territorial waters. The English captain and crew remained under arrest for many months before they were released and their prize returned to the French. The affair left the society's backers nearly bankrupt.

The most enduring legacy of the Anti-Gallicans is their remarkable map. It is striking in its full use of a large map sheet to convey cartographic information and in the absence of a typical decorative cartouche that might hint at French "frillery."

Beneath the title *A New and Accurate Map of the English Empire in North America* are two powerful statements. The first asserts the validity of British territorial claims: *Representing their Rightful Claim as confirmed by Charters and the formal Surrender of their Indian Friends.* The tone in the second statement changes from matter-of-fact to inflammatory: *Likewise the Encroachments of the French with the several Forts they have unjustly erected* characterizes the French as little more than a renegade nation. All of this is followed by the map's attribution, **by a Society of Anti-Gallicans**, whose Gothic script boldly implies English supremacy.

Vibrant color exaggerates the British colonial claims to the west while the French are relegated to a small area of Canada and their "encroachments," as signified in an anemic milky white. The color code is printed directly on the map in the sea just below Newfoundland: *The several Provinces of the English Empire in N. America are distinguished by Red, Blue, Green, and Yellow according to their respective jurisdictions. The Spanish by Brown; but the French Possessions and Encroachments are without any Colour.*

Two colored boundary lines running through the map are described in the ocean just off the Carolinas. *The French claim all the Country within the Hudson Bay Company's Southern Limits and the Brown Line. The Purple Line represents the Western Boundry of the hereditary & Conquer'd Country of our Indian Friends and Allies, which has been ceded and confirm'd to us by several Treaties and Deeds of Sale.* This establishes the territory east of the Mississippi and west of the Alleghenies as the theater of conflict between the "rightful" British claims and the French "encroachments."

The geography and color template in the main body of the Anti-Gallican are similar to several other contemporary maps but most of the side panel insets were, ironically, the work of a Frenchman, Jacques-Nicolas Bellin. The smaller insets accurately foreshadow the future points of conflict between Britain and France. Five show important Canadian ports, and both Louisbourg and Quebec would become sites of decisive engagements of the war. The map of Crown or Scalp Point anticipates the battles at Lake George and Ticonderoga for control of the inland waterway. By the time of the map's publication the British general Edward Braddock had already been defeated near Fort Duquesne, and the public would have a heightened interest in all of these potential flashpoints. The larger inset in the lower right emphasizes the Atlantic Ocean link between Europe and the North America colonies, critical to the English interest in trade and commerce. As historian Pedley observed, "By using easily accessible images to illustrate the borders of the map, the Anti-Gallican designer has emphasized sensitive and important areas of French territory, encouraging efforts against them."

The Anti-Gallican Society's membership and influence thrived during years of conflict with France but waned during periods of peace. Following the French and Indian War it became once again active when France supported the American colonies in the late 1770s and then during the Napoleonic Wars, after which the term "Anti-Gallican" faded into history save for a London hotel and pub that continued to use the name.

Braddock's March

In response to the French and Indian presence along England's colonial frontier, in 1754 the Duke of Cumberland appointed General Edward Braddock to command His Majesty's forces in North America. His choice was puzzling. Braddock was not a battlefield-tested officer. He had spent most of his military career on guard duty in London where he had ample opportunity to indulge his proclivities for drinking, gambling, and women. But Braddock had been an able administrator in Gibraltar and was a strict disciplinarian, a virtue valued by the soldierly duke. In anticipation of his departure for America, Braddock and Cumberland conferred over maps and planned the opening campaign–the capture of Fort Duquesne (figure 3) at the forks of the Ohio River, where the French had been most menacing. On January 13, 1755, Braddock sailed for Alexandria, Virginia, filled with confidence that his plan would be a success.

Braddock's confidence was fortified in large measure by the unreliable maps that greatly understated the distances in America and were not informative about the rugged topography of the land. James Fenimore Cooper noted "a feature peculiar to the colonial wars of North America, that the toils and dangers of the wilderness were to be encountered before the adverse hosts could meet." Braddock discovered that warfare in North America bore no resemblance to the choreographed battles that took place on the plains of Europe.

In 1753 Governor Robert Dinwiddie sent twenty-one-year-old George Washington to reconnoiter the Ohio country, where he found the French were constructing fortifications on land that had been claimed by the English. Washington returned the following year with a group of three hundred Virginia militia and Indian allies, and they overwhelmed a French patrol. Washington's chief ally on the mission, the Indian leader Tanacharison, killed the French commanding officer. That set in motion a large detachment of French and Indians who attacked the Virginians at their quickly constructed stockade, aptly named Fort Necessity. The little fort was soon filled with the dead and wounded and Washington was forced to accept the French commander's terms of surrender. As news of the defeat spread, colonial governors appealed for British intervention to unseat the French, and so, in the words of Horace Walpole, "The volley fired by a young Virginian in the backwoods of America set the world on fire."

Braddock's campaign was the starting point on the road to American independence. He reached Hampton, Virginia, on February 20, 1755, accompanied by the 44th and 48th Irish regiments (approximately two thousand men) under Colonels Sir Peter Halkett and Thomas Dunbar. Unloading the hardware of warfare was the first of many daunting tasks. A team of forty horses was required to pull each of the four 3,000-pound cannon from the ships. Braddock was not satisfied. He thought additional firepower was necessary and convinced Commodore Augustus Keppel to supply four additional 12-pounders from his flagship along with thirty seamen facile in the block and tackle lifting that would be required to transport the guns through the mountains.

Braddock's battle plan had two components. The first involved the army's movement up the Potomac River to Fort Cumberland, an outpost in the foothills of the Alleghenies. From there, a second contingent would proceed across the mountains to Fort Duquesne. The maps optimistically estimated these distances at approximately 30 miles and 15 miles, respectively. Colonel John St. Clair had sailed to America to prepare for the campaign, and he realized that the falls of the Potomac could not be blown up to make the river navigable. The army could not proceed by water and the distances would be more arduous and considerably farther than shown on the maps. Braddock sensed that Washington's knowledge of Virginia and the Ohio country could prove valuable and he asked him to join his "Military Family" as aide-de-camp.

The rugged terrain delayed the progress of the troops. For example, an advance army of three hundred ax men was necessary just to clear a twelve-foot roadway. After a two-month struggle in the wilderness the army reached Fort Cumberland short of horses, wagons, and supplies and on the verge of collapse. Fortunately in Frederick, Maryland, Braddock had met with Benjamin Franklin, who agreed to help resupply the army and also volunteered advice on fending off Indian ambuscades. "These savages may, indeed, be a formidable enemy to your raw American militia," Braddock had replied, "but upon the King's regular and disciplined troops, sir, it is impossible they should make any impression."

Once the army reached Fort Cumberland, it took another month to execute the final leg of the campaign. On June 8, Braddock wrote to London that the geographical information given him was "utterly false"; "Nothing can well be worse than the road I have already pass'd [120 miles] and I have a hundred and ten miles to march Thru an uninhabited

Figure 3. *Plan du Fort Duquesne*, cartographer unknown, 1755, manuscript, 49 × 33 cm. Bibliothèque nationale de France

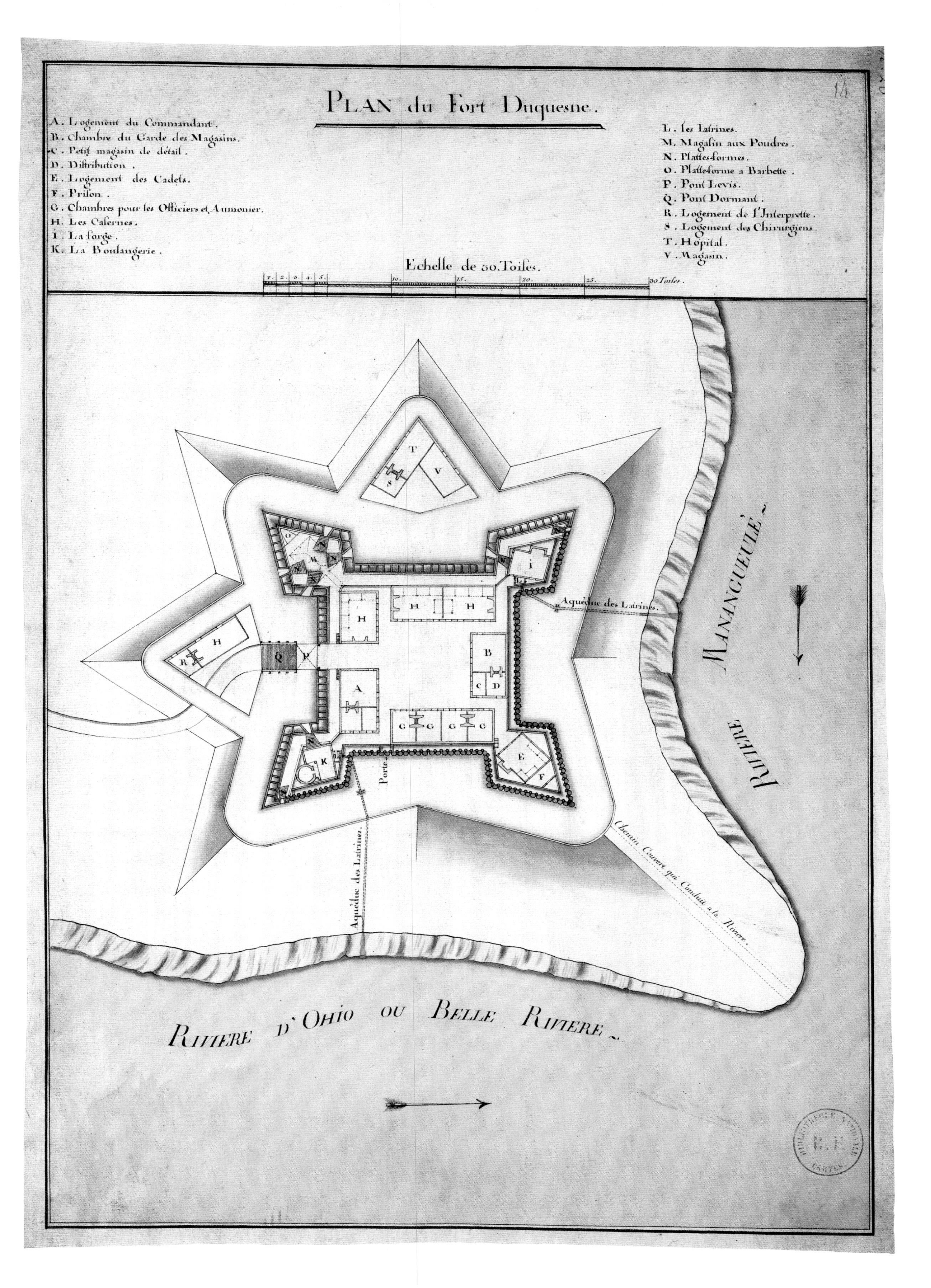

Plan du Fort Duquesne.
A. Logement du Commandant.
B. Chambre du Garde des Magasins.
C. Petit magasin de détail.
D. Distribution.
E. Logement des Cadets.
F. Prison.
G. Chambres pour les Officiers et Aumonier.
H. Les Casernes.
I. La forge.
K. La Boulangerie.
L. les Latrines.
M. Magasin aux Poudres.
N. Plattes-formes.
O. Platteforme a Barbette.
P. Pont Levis.
Q. Pont Dormant.
R. Logement de l'Interprette.
S. Logement des Chirurgiens.
T. Hôpital.
V. Magasin.
Echelle de 30 Toises.
Aquéduc des Latrines
Porte
Chemin Couvert qui Conduit a la Riviere
Riviere Manangueulé
Riviere d'Ohio ou Belle Riviere

wilderness over steep rocky mountains and almost impassable Morasses." Braddock ordered a "Working Party" to convert Washington's footpath over the Alleghenies into a proper roadway for the army with six hundred horses pulling artillery and provisions. The French governor Michel-Ange Duquesne boasted that "the English could not cross the Alleghenies in sufficient force to be a problem." Noted historian Lawrence Henry Gipson wrote, "I know of no other feat in the annals of the military history of North America that can be compared with it." The historian A. B. Hulbert would describe the mountain march as the "eighth wonder of the world."

Figure 4. *Captain Robert Orme*, Sir Joshua Reynolds (1723–1792), 1756, oil on canvas, 239 × 147 cm. © National Gallery, London/Art Resource, NY

Robert Orme (figure 4), Braddock's aide-de-camp, recorded the campaign in a journal and drew five remarkable maps, two of which are exhibited here. His first map (figure 5) identifies nineteen campsites en route to Fort Duquesne. Five days were required to cover the first twelve miles to Martin's Plantation (2). On the sixth day, Orme described the passage to Savage River (3) as a "rocky ascent of more than two miles, in many places extremely steep; its decent is very rugged and almost perpendicular; in passing which we entirely demolished three wagons and shattered several." The next day's route to Little Meadow (4) led through the dense pine forests called the Shades of Death, where, according to Colonel St. Clair, "a man might go 20 miles without seeing before him ten yards." From campsite to campsite the troops' anxiety increased, reaching a peak at Great Meadow (8, 9), where the bones of Washington's force from the previous year lay scattered.

As the army deliberately advanced, Braddock followed Washington's advice to create a "Flying Column" to push forward with thirteen hundred of the best men. Supplies and the remaining men under Colonel Dunbar followed in the rear. The column moved so quickly that confidence grew that the French and Indians could harass only their periphery. Orme noted, "They already permitted us to make many passes which might have been defended by a very few men." On July 8, the advance portion of the army reached the last campsite, the Monongahela (19). At this point, a decision was made to avoid a mountain pass and approach Fort Duquesne by two crossings of the Monongahela River, shown by the dotted line on Orme's map. The river was some three hundred yards wide and three feet deep with high banks on both sides, but an advanced guard under Colonel Thomas Gage crossed it unopposed. It was quite a sight, according to one diarist, "their bayonets fixed, Colors flying, and Drums and fifes beating and playing." Gage's advance force emerged from the river at Frasier's House, only seven miles from Fort Duquesne; the detail of the July 9 battle is shown on Orme's second map (figure 6).

Inside Fort Duquesne, Claude-Pierre Contrecoeur had barely a thousand men under his command and was considering capitulation when scouts reported the British were advancing along the vulnerable river route. Upon hearing this, a young Canadian officer in Indian dress, Daniel Beaujeau, rallied reluctant Indian warriors: "The English are going to throw themselves into the lion's mouth . . . hide yourselves in the ravines which line the roads, and when you hear us strike, strike yourselves. The victory is ours!"

Figure 5. *A Map of the Country between Wills Creek and Monongahela River*, in Robert Orme, *A Journal of the Expedition to North America in 1755, under General Braddock*, 1755, manuscript, 28 × 18 cm. © The British Library Board [Kings MS 212 (Part 1)]

The opposing forces caught a glimpse of each other two miles from the fort. Beaujeau (1) immediately directed a portion of approximately six hundred men into the ravines on both sides of the road and the others were ordered to take the high ground on the hill (S). Firing from behind dense cover they inflicted heavy casualties on Gage's advance party

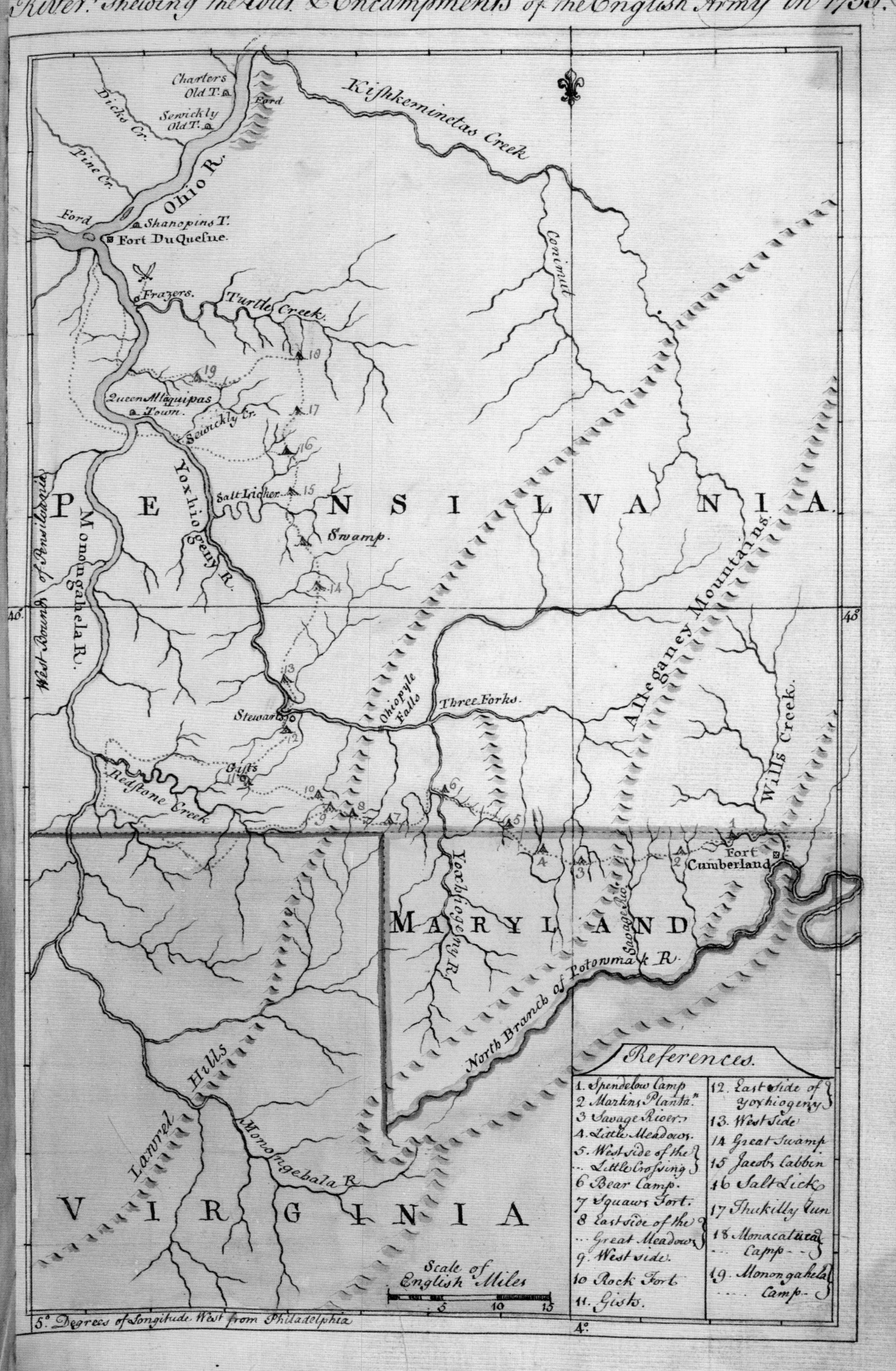
A Map of the Country between Wills Creek & Monongahela River: shewing the Rout & Encampments of the English Army in 1755.
Charters Old T.
Sewickly Old T.
Dicks Cr.
Pine Cr.
Ford
Ohio R.
Shanopins T.
Fort DuQuesne
Kiskeminetas Creek
Frazers
Turtle Creek
Queen Allaquipas Town
Sewickly Cr.
Yoxhiogeny R.
Salt Licks
Swamp
Continued
P E N S I L V A N I A
Monongahela R.
West Bounds of Pensilvania
Stewarts
Gists
Redstone Creek
Ohiopyle Falls
Three Forks
Allegany Mountains
Wills Creek
Fort Cumberland
M A R Y L A N D
Yoxhiogeny R.
Savage Riv.
North Branch of Potowmack R.
Laurel Hills
Monongehala R.
V I R G I N I A
Scale of English Miles
5 10 15
5° Degrees of Longitude West from Philadelphia
4°
References.
1. Spendelow Camp
2 Martins Planta.
3 Savage River
4. Little Meadows
5. West side of the Little Crossing
6 Bear Camp.
7 Squaws Fort.
8 East side of the Great Meadows
9. West side
10 Rock Fort
11. Gists.
12. East side of Yoxhiogeny
13. West side
14 Great Swamp
15 Jacobs Cabbin
16 Salt Lick
17 Thukilly Run
18 Monacatuca Camp
19. Monongahela Camp

A Plan of the Field of Battle & Disposition of the Troops, as they were on the March at the time of the Attack, on the 9th of July 1755.

Ohio R.

Fort du Quesne

Monongahela River

1.

S

Fraziers

Turtle Creek

Part of the Narrows

References.

1. French and Indians when discovered by the Guides

British Troops.

A. Guides with 6 Light Horse
B. Van of the Advanced Party
C. Advanced Party Commanded by Lieut.t Col. Gage
D. The Working Party Commd.d by Sr John St. Clair
E. Two field Pieces 6 Pounders
F. Guard to Ditto.
G. Tool Waggons.
H. Flank Guards.

Main Body of the Army

I. Light Horse
K. Sailors
L. Serj.ts & 10 Grenadiers
M. Subalterns & 20.
N. 12 Pounders.
O. Company of Grenadiers
P. Van Guard.
Q. Train of Artillery
R. Rear Guard of the whole Army
S. A Hill.
✗. Ground where the Principal part of the Battle was Fought

N B. The Distance from Fraziers House to Fort du Quesne is 7 Computed Miles

and the working party under Colonel St. Clair (A–H). Braddock responded by sending the main body (I–R) to assist Gage and help take the strategically important hill. Before these men could prepare for action, they were met head-on by the advance troops retreating in panic. Virtually the entire army was compressed in the 12-foot-wide roadway. The troops fired at an unseen enemy, killing many of their own ranks instead. Lieutenant William Dunbar recorded, "The officers . . . soon became the mark of the Enemy who scarce left one, that was not killed or wounded. Numbers ran away, nay fired on us, that would have forced them to rally." The site of the heaviest action is highlighted on Orme's map by the crossed swords.

"I cannot describe the horrors of that scene, no pen could do it. The yell of the Indians is fresh on my ear, and the terrific sound will haunt me until the day of my dissolution," lamented Lieutenant Matthew Leslie. On the field, Colonel Halkett was killed and Gage, Orme, and St. Clair were seriously wounded. Braddock had four horses shot out from under him, while attempting to rally his troops, before suffering a grievous wound of his own. Among the senior officers, only Washington remained unharmed, although several bullets had pierced his garments. As the army fell back, more soldiers were killed until they crossed the river, where the Indians halted, preferring to return for their "trophies," or scalps, before nightfall. British casualty rates (killed and wounded) were over 70 percent, a higher rate than sustained at the Charge of the Light Brigade a century later.

A sledge carried the mortally wounded Braddock back to Colonel Dunbar's encampment. "Who would have thought it?" Braddock said to Orme. The next day, he murmured, "We shall better know how to deal with them another time," and he died soon after. As the ranking surviving officer, Dunbar assumed command but he had no taste for renewing the battle. He destroyed and buried the ordnance and excess stores and beat a hasty retreat to Philadelphia, leaving the entire frontier in the hands of the French and their Indian allies.

The Battle of Lake George

At the cusp of America's great military engagements of the eighteenth century stands the Battle of Lake George on September 8, 1755, viewed by many as a seminal event on the road to American independence. The historian Elizabeth Seelye characterized it as "the first struggle in which the untrained farmers of America faced the finely drilled troops of the Old World and learned the courage which afterwards led them to dare to bring on the struggle of the Revolution." In addition to its military significance, this was the first battle of the French and Indian War to be recorded by American mapmakers, and two remarkable colonial maps delineate the action of that late summer day.

Figure 6. *A Plan of the Field of Battle & Disposition of the Troops*, In Robert Orme, *A Journal of the Expedition to North America in 1755, under General Braddock*, 1755, manuscript, 28 × 18 cm. © The British Library Board [Kings MS 212 (Part 5)]

The two combatants in the field came from vastly different backgrounds. The French had captured Braddock's exhaustive campaign orders near Fort Duquesne in July 1755, and the French government in Montreal directed the Baron de Dieskau to oppose any British military incursions north of Albany. Dieskau, commander of all French troops in North America, was a sophisticated soldier who had already served in Europe's finest army, ultimately being promoted to the rank of major general.

William Johnson, in command of the American troops, on the other hand, was a man with no military experience. Braddock had appointed him at the conference of colonial governors in Alexandria, Virginia, largely because of his knowledge of the Indians; he had lived among the Mohawks and was thoroughly acquainted with their customs and language. Johnson also had the advantage of being one of the richest men in America, having acquired vast holdings in land. His first assignment was to raise four thousand volunteers from New York and New England, supplemented by Mohawk warriors, and then attack Fort St. Frederick (Crown Point). The French used this fort as a base for raids, and it had become "a Sharp Thorne" in the sides of the colonists of New York and New England.

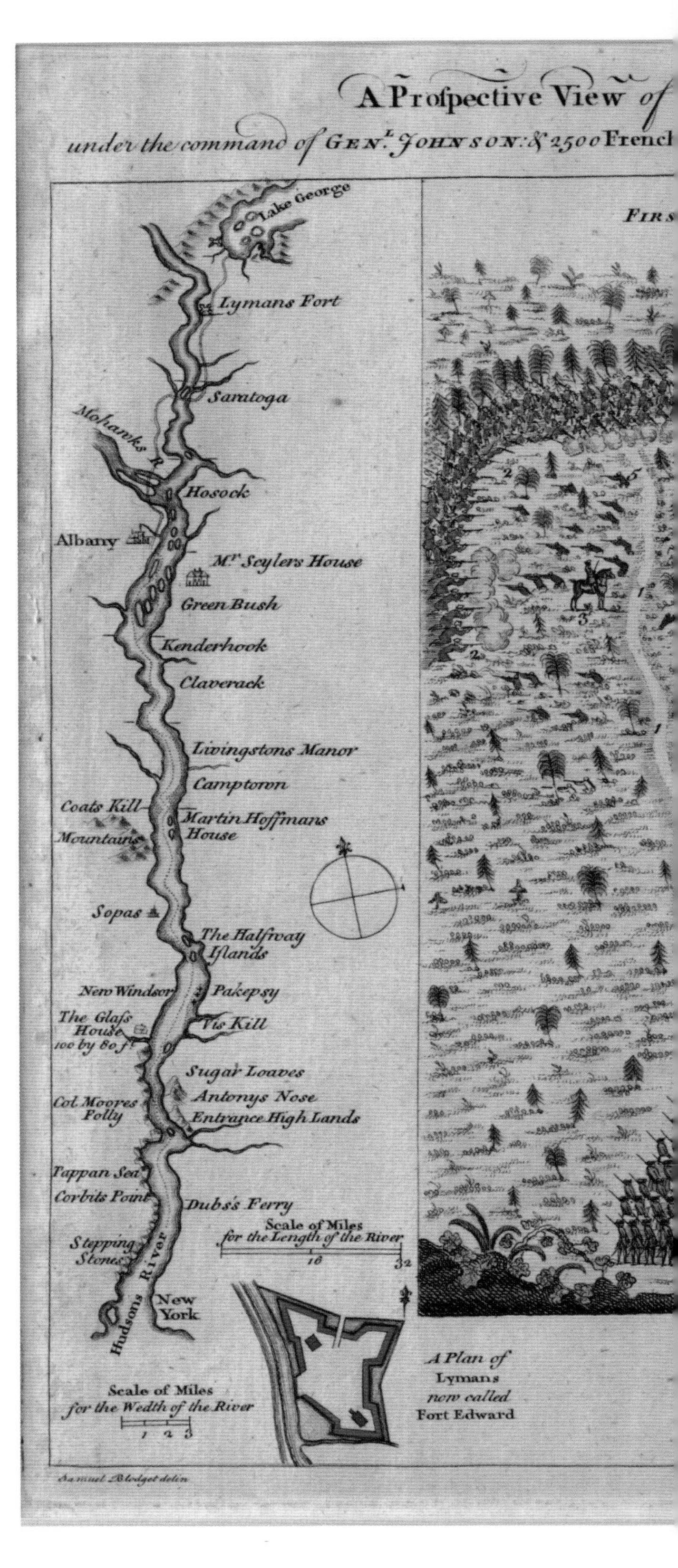

Johnson's ragtag army left Albany late in the summer of 1755, traveling up the Hudson River to Fort Edward and then marching overland to encamp on the south shore of Lake George. When Indian scouts reported on September 8 that Dieskau was nearby with three hundred French regulars and some twelve hundred Canadian militia and Indians, Johnson boldly ordered Colonel Ephraim Williams of Massachusetts and the Indian chief Hendrick to confront the French general with an equivalent force. After two hours of fighting, Williams and Hendrick lay dead on the battlefield. Fortunately for higher education in America, the unmarried Williams had drawn a will, leaving most of his fortune to the establishment of Williams College, named in his honor.

Dieskau pursued his attackers. As gunshots grew closer to Johnson's camp, orders were given to overturn wagons and position three cannon to cover the road. Dieskau soon reached the clearing outside the camp and victory appeared at hand when the Indians and Canadians balked at the sight of the cannon and sought positions behind trees and bushes. As General Johnson's secretary recorded, "This happy Halt, in all Probability saved Us, or

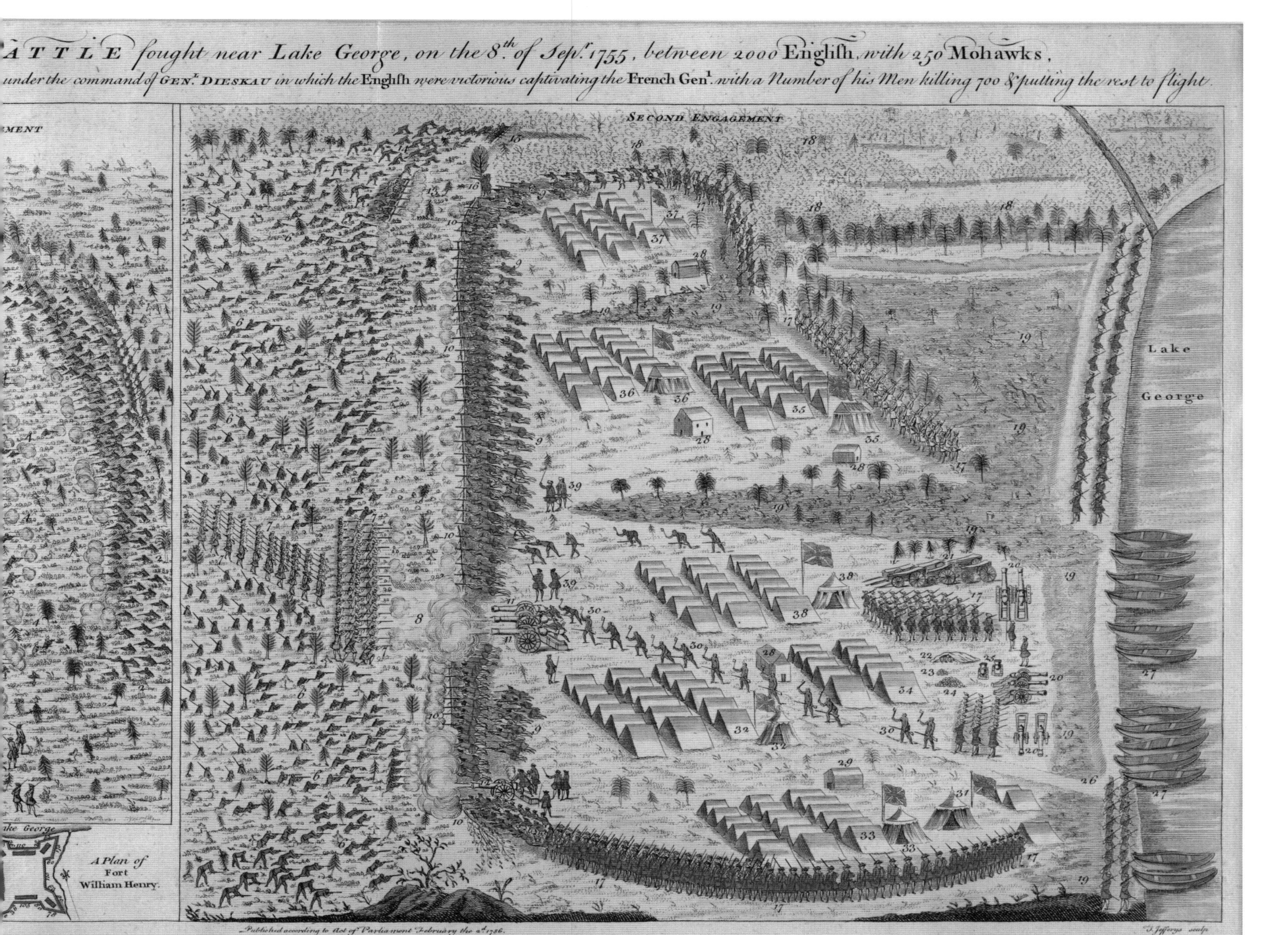

the French General would have continued his pursuit and I am afraid would have entered in with the last of our flying Men, before our troops recovered their consternation."

Dieskau then ordered his two elite French regiments to march on the camp. When they had advanced halfway, provincial guns cut "Lanes, Streets and Alleys" through them, virtually annihilating both regiments. An intense exchange of gunfire continued for several hours. "It was the most awful day my eyes ever beheld," wrote the surgeon Thomas Williams, Ephraim's brother, to his wife; "there seemed to be nothing but thunder and lightning and pillars of smoke." By midafternoon, the French forces began a disorderly retreat without Dieskau, who was seriously wounded and propped up against a tree. Seething from the loss

Figure 7. *A Prospective View of the Battle Fought near Lake George, on the 8th of Sepr. 1755*, Samuel Blodget, London, 1756, copperplate engraving, 26 × 51 cm. Private collection

Figure 8. *The Brave Old Hendrick the Great Sachem or Chief of the Mohawk Indians*, Eliz. Bakewell, London, ca. 1740, copperplate engraving, 32 × 24 cm. Courtesy of the John Carter Brown Library at Brown University

of several chiefs, the Mohawks closed in to kill him but Johnson, himself wounded, intervened. Dieskau wrote that he owed his life to Sir William Johnson without whose interference "I should have been assuredly burned at a slow fire by the Iroquois."

There was no relief for the French forces. When they retreated to the place where Williams had been killed in the morning, two hundred provincials sent out from Fort Edward ambushed them. The surviving members of Dieskau's army fled, leaving behind baggage that contained important documents.

Samuel Blodget, an artist who essentially served as a war correspondent, recorded the vivid events. "An Independent Person, not belonging to the Army, I had, it may be, as good an Opportunity, as any person whatever, to observe the whole management of both sides." Blodget made a drawing of the action he saw and then transported it to Boston. There he persuaded Thomas Johnston, arguably the finest engraver in colonial America, to print the scene.

Samuel Blodget's *Prospective Plan of the Battle Fought near Lake George on the 8th of September 1755* was published on December 31, 1755. Although crude by European standards, Thomas Jeffreys, London's leading map publisher, immediately reprinted it in early 1756, declaring it "the only piece that exhibits the American method of bush fighting." The few surviving copies of Blodget's Boston printing, with a slightly different title, are darkened with age; figure 7 shows a colored example of the English edition.

Blodget uses numerals to locate key events of the battle, which are explained in a five-page supplement printed with the map. In the first engagement, (1) shows the road along which Williams sought out the enemy while (2) depicts the Canadians and Indians lying in wait for the ambush (in the form of a hook). Chief Hendrick (figure 8) is shown in (3) "on horseback because he only could not well travel on foot, being somewhat corpulent as well as old," while in (4) and (5) the provincials and their Indian allies start to break rank and begin their retreat.

The second engagement details the attack on the camp. The Canadians and Indians advance

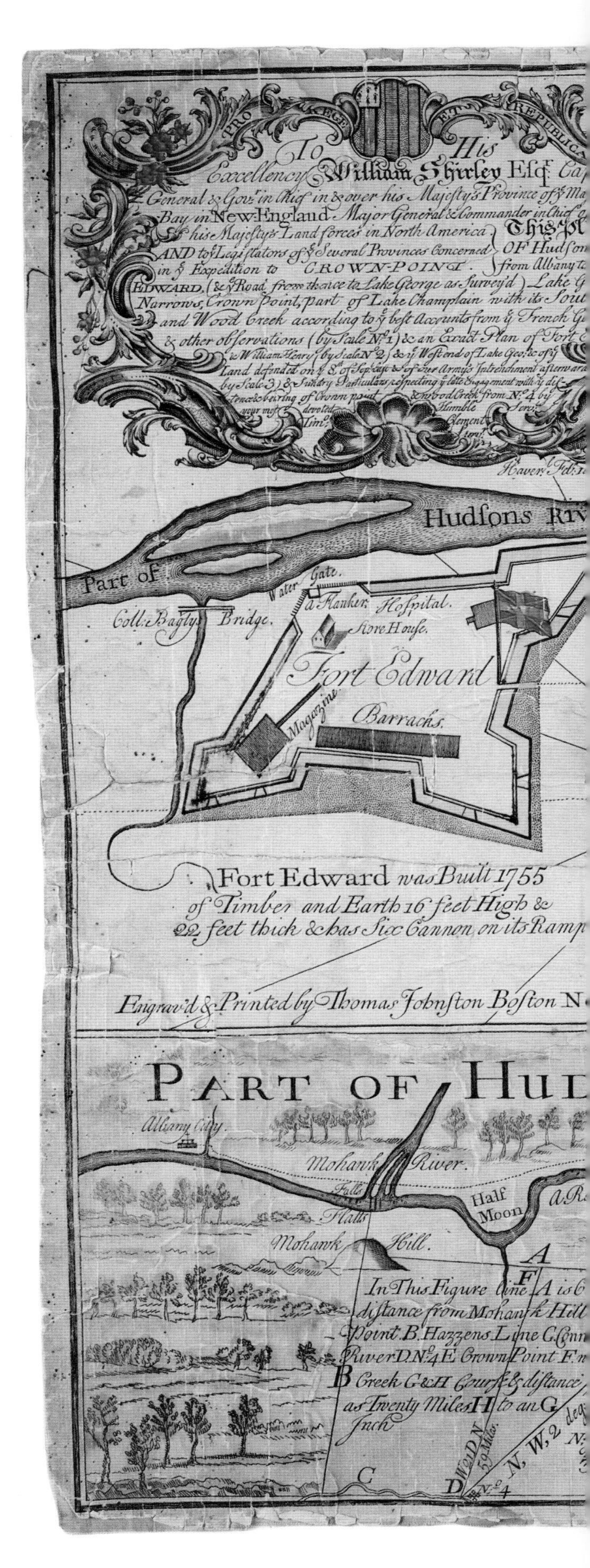

Figure 9. *This Plan of Hudson's Rivr: From Albany to Fort Edward, (& Ye Road from thence to Lake George as Survey'd)*, Timothy Clement, Boston, 1756, copperplate engraving, 44 × 67 cm. Courtesy American Antiquarian Society

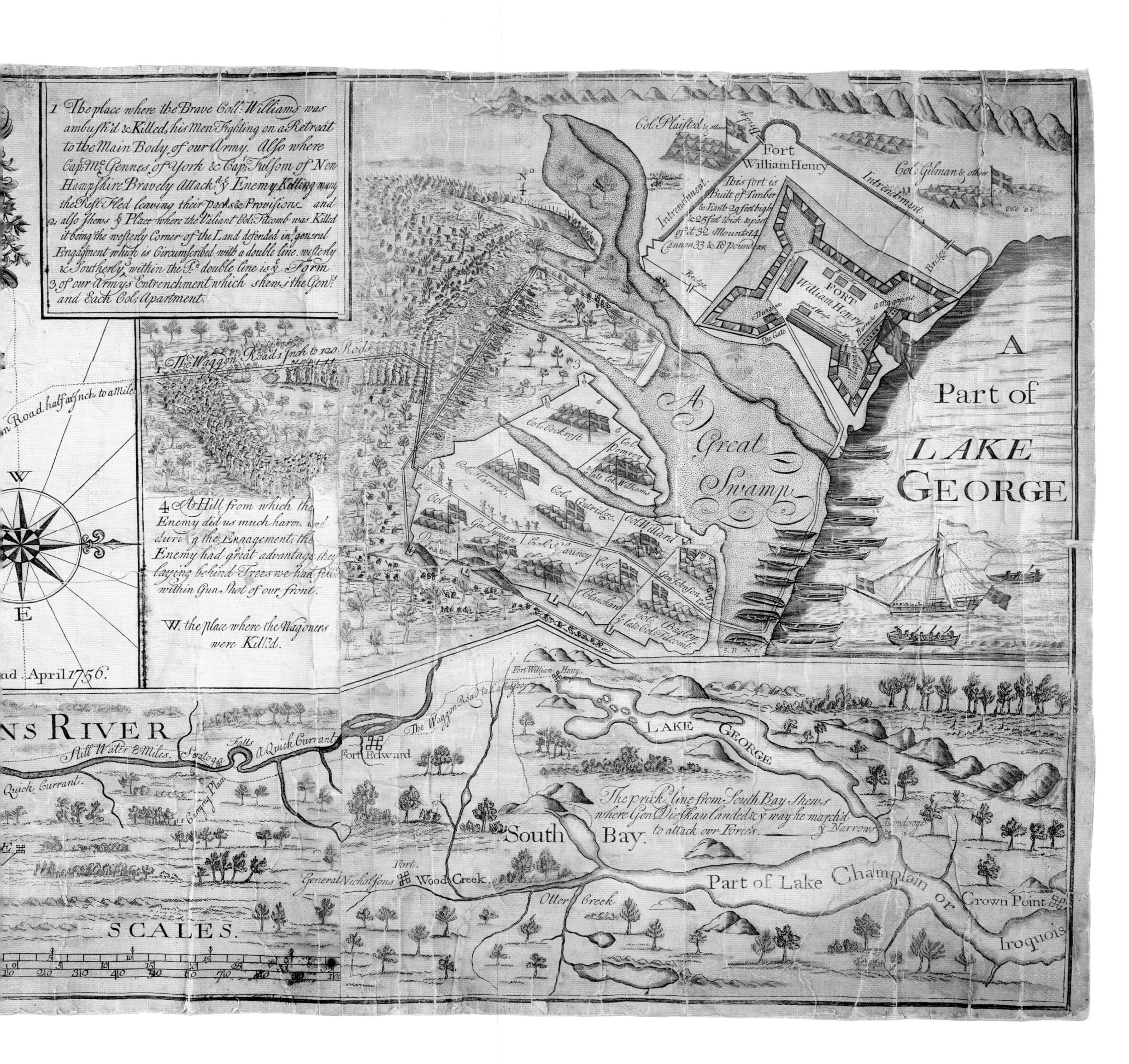

1 The place where the Brave Coll: Williams was ambush'd & Killed, his Men Fighting on a Retreat to the Main Body of our Army. Also where Capt. McGennes of York & Capt. Fulsom of New Hampshire Bravely Attack'd y Enemy Killing many the Rest Fled leaving their packs & Provisions and
2 also shews y Place where the Valiant Col: Titcomb was Killed it being the westerly Corner of the Land defended in y general Engagment which is Circumscribed with a double line. westerly & Southerly within the sd double line is y Form
3 of our Army's Entrenchment which shews the Genl. and Each Col: Apartment.
4 A Hill from which the Enemy did us much harm during the Engagement; the Enemy had great advantage they laying behind Trees we had felled within Gun Shot of our front.
W. the place where the Wagoners were Kill'd.
The Waggon Road 1 Inch to 120 Rods
Road half an Inch to a Mile
W
E
April 1756.
Col: Plaisted & others
Fort William Henry
Col: Gilman & others
Intrenchment.
This fort is Built of Timber & Earth 20 feet high & 25 feet thick & part of it 32 Mount 14 Cannon 33 & 18 pounders
Bridge
FORT William Henry
The Gate
Barracks
N° 4
Col: Cockroft
Col: Pomroy
late Col: Williams
Col: Harriss
Col: Gutridge
Col: Willard
Col.
Gen: Lyman
Col: Blanchard
Gen: Johnson
Col: Bagley
late Col: Titcomb
A Great Swamp
A Part of LAKE GEORGE
NS RIVER
Still Water 8 Miles
Saratoga
Falls
A Quick Currant
Quick Currant
Carrying Place
The Waggon Road to Lake Geo
Fort William Henry
Fort Edward
LAKE GEORGE
The prick'd line from South Bay Shews where Gen: Dieskau Landed & y way he march'd to attack our Forces.
South Bay.
y Narrows
Ticonderoga
Fort
General Nicholsons
Wood Creek
Otter Creek
Part of Lake Champlain or Iroquois
Crown Point
SCALES.

around the perimeter (6) as the French regulars approach in their disciplined formation (7) along the cleared road (8). The form in which the provincials were fighting is illustrated in (9), behind a breastwork of logs (10). The three heavy cannon that decimated the French regulars appear in (11). Blodget continues with another 28 references, providing candid views of the conditions within the camp.

Blodget's *Perspective View* had been in circulation for merely a month when another remarkable map appeared: *This Plan of Hudson's Rivr: From Albany to Fort Edward, (& Ye Road from thence to Lake George as Survey'd)*, dated February 10, 1756 (figure 9). Its author, Timothy Clement, was also at Johnson's camp, serving as a surveyor, soldier, and scout. Clement and Blodget may have collaborated on their works as both lived in Haverhill, Massachusetts. Thomas Johnston was again the engraver when it was printed in April 1756.

The significance of Clement's map lies in the inset printed at the bottom that documents the path that Johnson's army took from Albany. Lakes George and Champlain and the French forts at Ticonderoga and Crown Point are also delineated. This new information came from documents found in Dieskau's captured baggage. Clement successfully petitioned the State of Massachusetts for compensation, stating, "I have here Gentlemen Laid to your view a Plan of what I surveyed, with sum farther intelligence of the Laying of the Lakes and of their Distance: Which I took from the French Plans: Which we took on the 8th of Septr. from the French Generall."

Only one complete copy of Clement's map was known before 1944, when the American Antiquarian Society acquired the example illustrated here for $400 (likely the highest price paid to that date for a map published in America). In the October 1944 society report the map was heralded as one of the "best known of the Library's possessions . . . a fresh clean copy with contemporary coloring, . . . [owing] its unique condition to the fact that it was an officer's field copy, mounted on cotton twill and rolled around a rod." The map's importance was significantly enhanced by an accompanying document that stated Sir William Johnson had given this "proof" copy of the map to his superior the British general Robert Monckton.

Albany

On September 22, 1609, a flat-bottomed ship, the *Half Moon*, lay at anchor 150 miles north of the New York harbor. Brightly painted with geometric designs and festooned with the colorful flags of the Dutch East India Company, the ship heralded a celebration that would never take place. Disheartening news arrived the next morning from ship's mate Robert Juet who reported that crew members exploring upriver had found but "seven foot water, and unconstant soundings." Henry Hudson's dream of a passage to the Pacific was dashed, but he had discovered the deepest navigable point into the colonial interior on the river that would bear his name. On the adjacent banks, Fort Nassau was built and renamed Albany after the Dutch surrendered it to the British in 1664.

pages 24–25: Figure 10. *Plan, of the City, of Albany, in the Province, of, New, York*, Thomas Sowers, 1756, manuscript, 44 × 89 cm. © The British Library Board (Maps K. Top 121.41)

The location of Albany provided unparalleled access to strategic waterways leading north to Lake Champlain and Canada, west to Lake Ontario, and south to the Atlantic Ocean. The town thrived as a trading center but during the colonial wars the waterways became the "Warpaths of Nations," and Albany found itself at the center of conflict. Early in 1756 Thomas Sowers, a British engineer, executed the manuscript map of Albany illustrated here (figure 10). It carefully depicts military and civilian sites and roadways in the town of two thousand inhabitants, populated in addition with ten thousand soldiers. Albany would be the headquarters for the British throughout the French and Indian War.

In the mid-eighteenth century Albany was a prosperous Dutch city of handsome stone buildings and broad streets. The Albanians, as they were then called, shipped furs, lumber, and grain to New York City, with rum the principal product received in return. Peter Kalm, the Swedish naturalist and explorer, who visited the city in 1749, was alarmed by the "avarice, selfishness and immeasurable love of money of the inhabitants of Albany." They plied the Indians with rum to "cheat [them] in the fur trade." This was a dangerous game, and although the nearby Mohawks typically allied themselves with colonial interests, the northern tribes of the Five Nations were often hostile. In 1745, the citizens of Albany were thrown into panic when six hundred French and Indians burned nearby Saratoga and killed or captured virtually all the town's inhabitants. Albany's Fort Frederick, (A) on the Sowers map, was vulnerable from nearby hills and, as Kalm observed, "can serve only to keep off plundering parties without being able to sustain a siege."

Figure 11. Benjamin Franklin's *Join, or Die*, 1754, woodcut, Photos, Prints and Drawings Division, Library of Congress

By 1754 French territorial encroachments had become such a concern that colonial and Indian delegates met in Albany to discuss a common defense. The Albany Congress took place at the Town House (I), and the scope of the proceedings was broadened considerably when delegate Benjamin Franklin proposed the Albany Plan of Union. This early attempt at unification was supported by the delegates but, not surprisingly, was frowned upon by the British Parliament. Franklin's "Join or Die" (figure 11) is a cartographic image emphasizing the importance of union. It was first published in newspapers in 1754 to encourage the colonies to unite in support of Britain against the French but by 1775 the image would become the symbol of colonial rebellion against the British.

On November 23, 1755, Colonel Thomas Gage wrote his former comrade in arms George Washington that "The two regts are camped here in a miserable situation." The 44th and 48th British regiments had spent nine grueling months under General Braddock in reaching Fort Duquesne only to be routed by the French in a matter of hours. The survivors retreated to Philadelphia and were then transported to dull and frigid winter quarters in Albany. Thomas Williams, the Massachusetts surgeon, reported, "A grievous sickness among the troops, we bury five or six a day. Not more than two thirds of our army fit for duty." "Long encampments are the bane of men," he concluded.

Sowers's map shows the encampment of Braddock's army (Q) and the regiments (R) formerly under Colonels Halkett and Dunbar. Braddock and Halkett were both killed at the Monongahela and Dunbar was relieved of command following his hasty retreat from the battlefield. Nearby are the independent companies (S), consisting of provincial soldiers incorporated directly into the British army for their effectiveness in irregular fighting

PLAN, of the, CITY, of, ALBAN
References
A. Fort Fredrick
B. Hospital and Barracks
C. The Governors House
D. The Kings Chaple
E. Brracks for ye British Troops
F. Magazine for Provisions
G. Market Houses
H. The Dutch Church
I. The Town House
K. Barrack for ye Artillery
L. The Town guard House
M. North
N. East
O. West
P. South
Bastions
Q. The Encampment of the British Troops that were Commanded by Major General Braddock
R. Sir Peter Halkett and Colonel Dunbars Regts
S. The two Indipendnt Companies
T. The Artillery
U. Powder Magazins
W. The new England Encampment
X. The Old Fort
Y. The Ferry House
Z. The Corn Mills
Thos Sowers Engr 1756
Road to Shonechady
Road to new York
HUDSON
N
Scale of 400 Feet to one Inch
0
400
800
1200
1600
2000

the PROVINCE of NEW YORK.

WOODS

Road to Nestagaunaugh

The Maner House of Renslearwyck

L

Road to ye Mills

W

W

The Road to Fort Edward

S RIVER

Figure 12. Silver salver presented to Captain Thomas Sowers, 1773, Collection of the New-York Historical Society

tactics. The New England colonials (W) were camped well north of the town along the river, highlighting the fact that colonial and British troops rarely got along well together.

The Sowers map conveys the isolation of the town bordered on one side by the river and on the other by the denseness of the woods. Sowers includes artistic touches that are not employed on many manuscript military maps, including the ornate rococo bordered cartouche and the tromp l'oeil ribbon banner bearing the title of the map. In early 1756 Sowers was transferred to Oswego, New York, on Lake Ontario where the newly appointed French commander Louis-Joseph de Montcalm defeated the British. Sowers served as chief engineer of a British force that retook Oswego in 1758. The following year he designed and oversaw the construction of Fort Ontario, whose outer walls still exist today.

Following the war, Sowers remained active as an engineer in colonial America. In 1773, a year before his death, he was given a large engraved silver plate by Governor William Tryon for his service in rebuilding the battery at the southern tip of Manhattan Island, a critical point of colonial defense. This "salver" is now one of the most prized treasures of the New-York Historical Society–"embellished with extraordinary engraving that depicts the seal of the City of New York" under which are gathered all the tools of the mapmaker's trade (figure 12).

During the Revolution, Albany would again become the pivotal military post on the continent's interior, with the Americans retaining control of the town in spite of determined British efforts to capture it. Following the war, Albany's dominance in trade continued into the nineteenth century thanks to the building of the Erie Canal. Originating at Albany and extending 363 miles to Buffalo, the canal opened the west to immigration and initially reduced shipping costs by 90 percent. It captured America's imagination to an extent that not even the transcontinental railroad nearly half a century later would equal. At the canal's opening in 1825, New York's governor DeWitt Clinton boarded the *Seneca Chief* in Buffalo carrying a bucket of Lake Erie water to be symbolically married with the Atlantic Ocean. The world's fastest communication line had been set up to announce Clinton's departure. Cannon strategically placed within earshot of one another conveyed the news to New York City in the then staggeringly short time of eighty-eight minutes.

Louisbourg

On June 1, 1758, Pierre-Jerome Lartigue watched the largest expeditionary force ever launched by the British sail into Gabarus Bay off Canada's Cape Breton Island. It consisted of 160 ships, 2,000 cannon, and 27,000 combatants. The fortress of Louisbourg, located on the rocky shores of the bay, was about to become the site of the most pivotal battle of the Seven Years' War. During the next fifty-four days, Lartigue would view the historic action from his house within the fortress and record his observations in the map pictured in figure 13.

The Lartigue manuscript map was completely unknown until its appearance at auction in 2013. It provides a view of the entire theater of battle and its uniqueness is enhanced by the fact that a Canadian executed it thirty-three years before any map was published in Canada. Unusual as the Lartigue map is, an anonymous manuscript map (figure 14), recently conserved at the Library of Congress, may also be the work of a Canadian. It is one of the most unusual and dramatic cartographic works of the entire war.

Detail, figure 14, pp. 30–31

France ceded Acadia (Nova Scotia) to the British at the 1713 Treaty of Utrecht but retained Cape Breton Island where a veritable citadel was constructed, named after the king of France. The buildings at Louisbourg were so enormous that Louis XV purportedly said he expected to see their peaks from Versailles. Four thousand French soldiers were garrisoned there in 1758, along with a like number of civilian residents. It was the third busiest port in North America and served as the French naval base for patrolling the Cabot Strait and Saint Lawrence waterway, the main artery into the interior of North America as seen on the Anti-Gallican map detail (page 33). From Quebec, the Great Lakes and the Ohio and Mississippi Rivers could be reached, providing the French with the freedom to establish forts and claim territories north and west of the Allegheny Mountains.

With only one tenth of the population of the British colonies, New France had nevertheless won a series of victories over its longtime enemy. "It was over England that the clouds hung dense and black," wrote the historian Francis Parkman. "Her prospects were the gloomiest." William Pitt, the leading member of the British Parliament in 1757, wrote, "We have lost all the waters. We have not a boat on the lakes. Every door is open to France." A bold attack was needed to break the French hammerlock on the waterways of North America. The Louisbourg campaign was led by Jeffrey Amherst and James Wolfe and the fleet by Admiral Edward Boscawen, whom Pitt praised by comparing him to lesser officers: "They always raise difficulties, you always find expedients."

The great engineer Sébastien Le Prestre de Vauban, who was famous for his sturdy military designs, inspired the fortress of Louisbourg. The main walls were over 30 feet high, protected by wide trenches and moats. Strategically placed cannon could repel attacks from both land and sea. The fortress occupied nearly a hundred acres and was situated in a natural harbor whose narrow entrance was protected by the island and lighthouse batteries–Batterie de l'isle and Tour de la lenterne on Lartigue's map. Inside the harbor, the 42-pound cannon of the Batterie Royale and French men-of-war lying at anchor provided further protection.

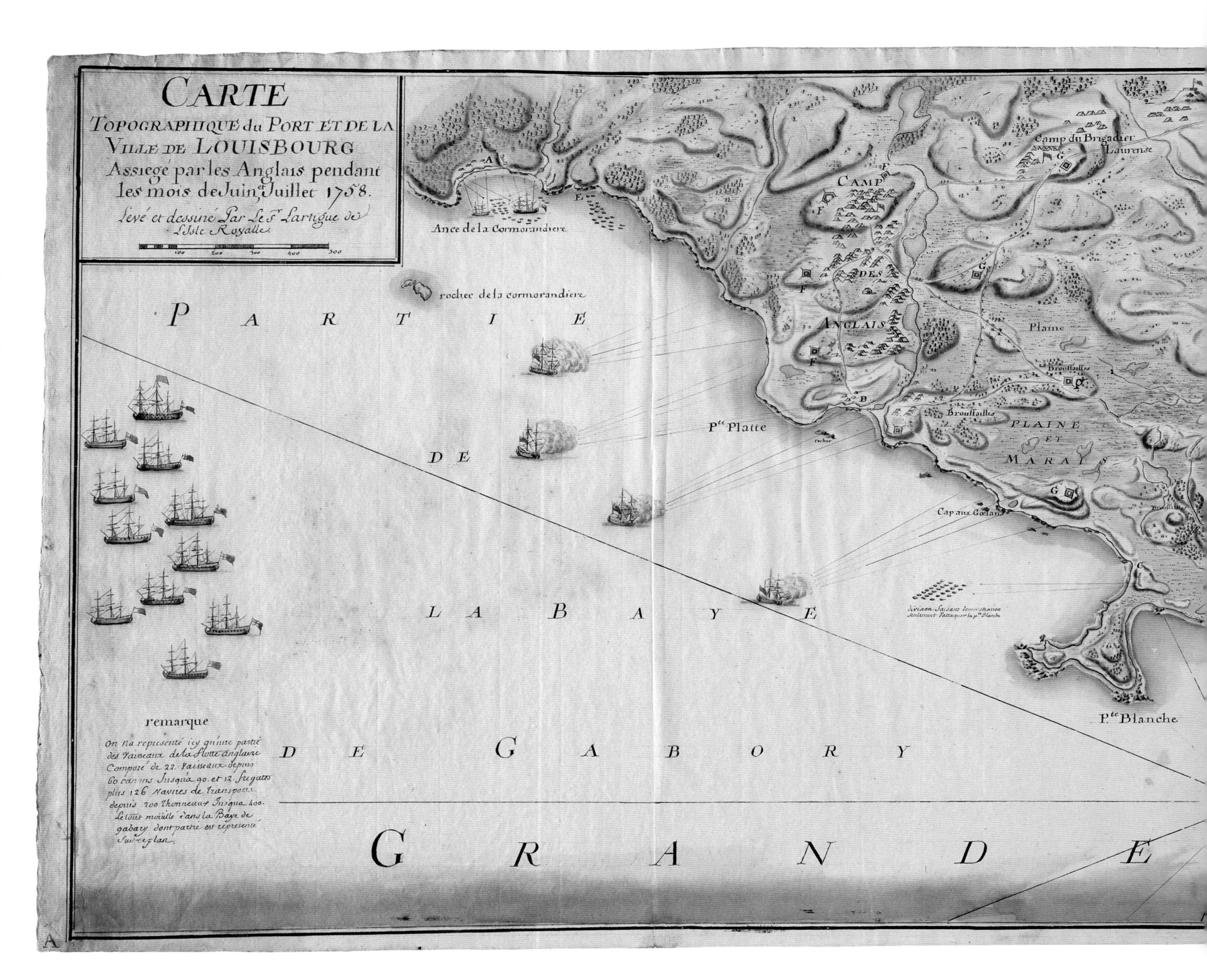

Figure 13. *Carte Topographique du Port et de la Ville de Louisbourg*, Pierre-Jerome Lartigue, 1758, manuscript, 40 × 105 cm. Private collection

It was not for nothing that Louisbourg became known as the "Dunkirk of North America." Amherst ruled out a direct attack and focused on the coasts, heavily guarded by French troops and artillery entrenched atop shoreline cliffs. He assigned Wolfe to lead an amphibious assault in small craft resembling whaleboats. Such a suicidal attack must have put fear in the hearts of the soldiers. Most could not swim.

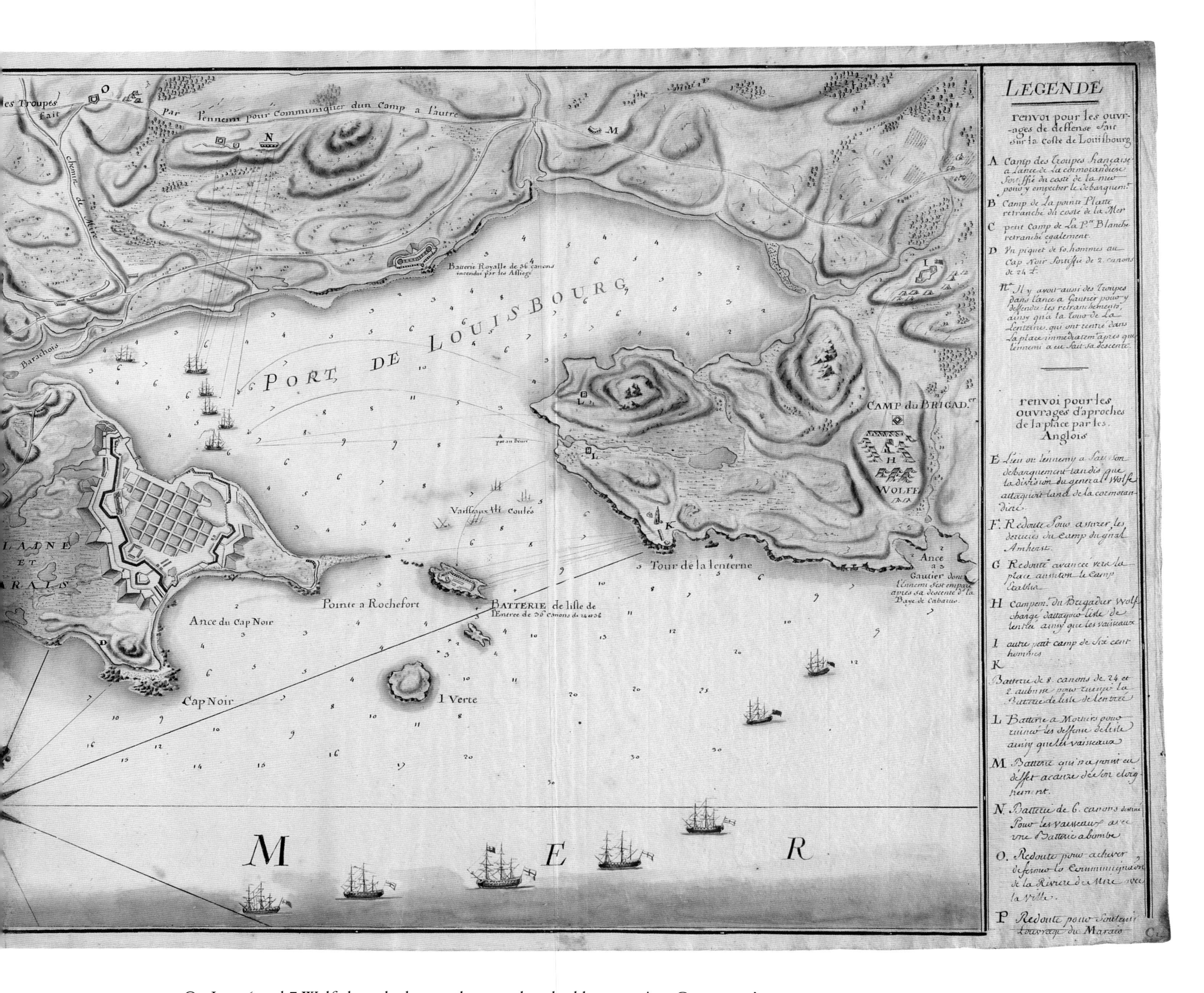

On June 6 and 7 Wolfe launched more than one hundred boats against Cormorant's cove–Ance de la Cormorandiere (A)–but heavy surf and deadly French fire prevented him from landing. On June 8 he called for a retreat but several boats managed to reach a protected inlet on the shore. Observing this from a distance inspired Wolfe and he made his way to the inlet and then commanded others to follow. Many drowned in the 40-degree

Figure 14. *Plan du Cap Breton, dit Louisbourg avec ces environs*, cartographer unknown, 1758, manuscript, 51 × 203 cm. Geography and Map Division, Library of Congress

water but those who survived were led up the cliffs and ordered to charge with bayonets. The astonished French retreated into the fortress.

"We made a rash and ill advised attempt to land, and by the greatest of good fortune imaginable we succeeded," Wolfe wrote to his father. Pitt's biographer Von Ruville described this as "the critical moment to which one can almost point as marking a change in the whole colonial war."

In the weeks that followed, the English brought their entire army and cannon ashore and established Camp des Anglais (F). Wolfe led his troops to the other side of the harbor, passing the Batterie Royale, destroyed by the French so its guns could not be turned upon the fortress. Wolfe then camped at (H) and positioned cannon at (K) from which he was able to destroy the lighthouse and island batteries. French ships still guarded the harbor entrance but on the night of July 21 a shell struck the magazine of the man-of-war *Celebere* and then the *Entreprenant* and *Capricieux* caught fire from the *Celebere*'s burning embers. The ships lit up the harbor throughout the night. Boscawen was soon free to enter the harbor and the bombardment from both land and sea became insufferable. "Not a house in the whole

place but has felt the force of their cannonade. Between yesterday morning and seven o'clock tonight a thousand to 1200 shells have fallen inside the town," wrote a French diarist.

On July 26 Governor Augustin Drucour sought honorable terms of surrender. His valiant defense of Louisbourg entitled his soldiers to parole according to eighteenth-century military etiquette. The British denied this request, citing French and Indian atrocities at Fort William Henry, and sent French troops to England as prisoners of war. Drucour's only satisfaction was that his defense had thwarted a follow-up attack against Quebec.

Amherst stationed a garrison at Louisbourg and then moved his army to Albany. Wolfe wrote to Amherst before he returned to England: "If you will attempt to cut up New France by the roots, I will come back with pleasure to assist." The next year, Wolfe would be immortalized at the Battle of Quebec.

Lartigue came from one of the oldest and most respected families in Louisbourg and served for a time as the king's Keeper of Stores. He likely refined his cartographic skills while living in Paris from 1745 to 1749 as his map exhibits the French style of the time. In 1754 Lartigue wrote Le Courtois de Surlaville that it is "too bad that he had lived for such a long

Le Capricieux de 64 canon
Brulle
Le prudent comandant de 74 canon
Brulle par les anglois
L'entreprenant
de 74 canon brulle
le Selebre Brulée de 64 canon
Etant
port de la reine
Cap Noir

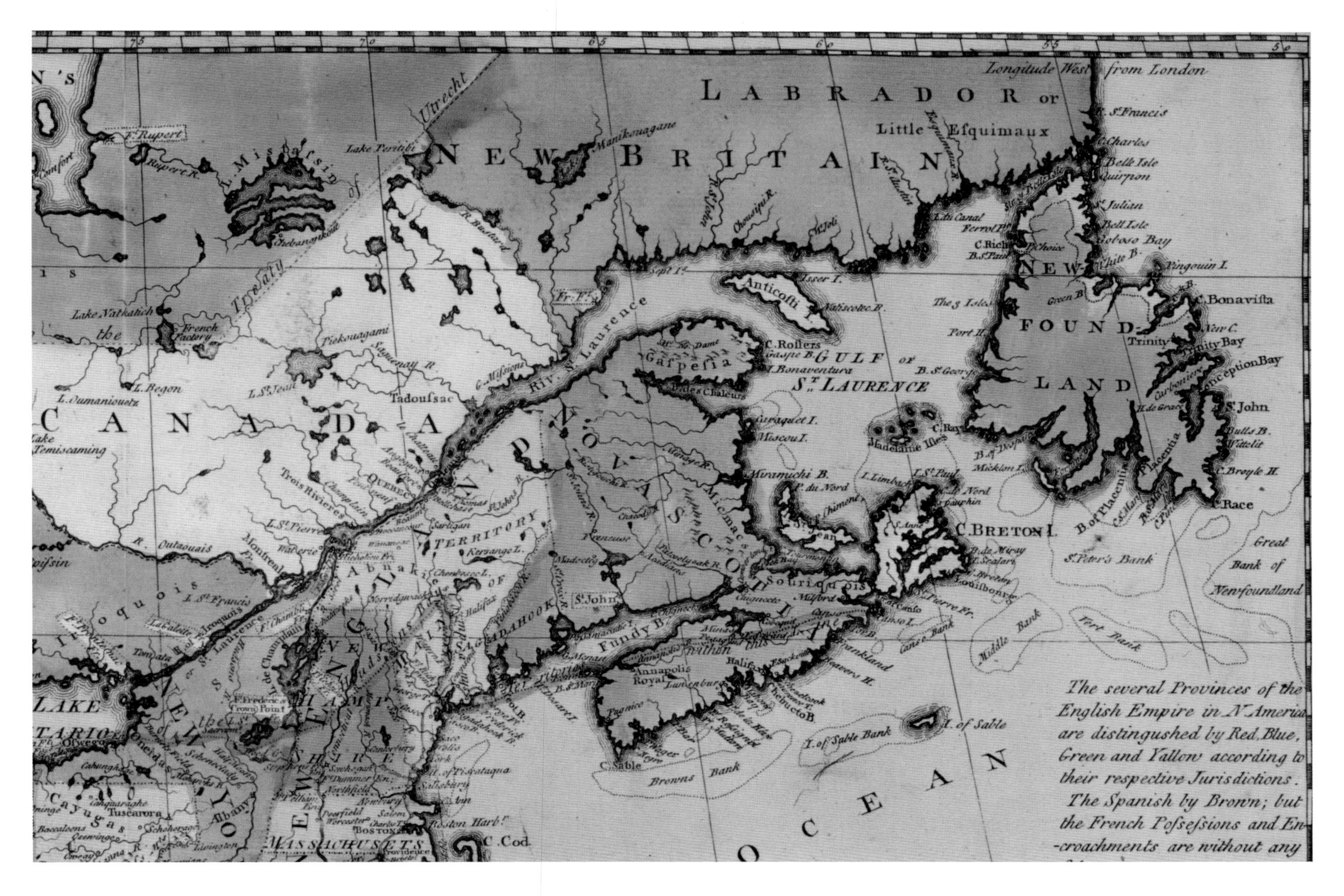

Detail, figure 1, *A New and Accurate Map of the English Empire in North America*, Society of Anti-Gallicans, London, 1755, pp. 8–9

time in a country so rugged, and where the men are so little civilized." Neither Lartigue nor anyone else would live in Louisbourg after 1760. That year, the British dug fifty tunnels, filled them with explosives, and demolished the fortress. As historian Lawrence Henry Gipson observed, "Thus this once flourishing city, void of inhabitants and wrecked beyond repair, took its place among other great ruins of the past." Lartigue and his family moved to French Guiana where he died in 1772.

It is probable that a Canadian also created the manuscript map at the Library of Congress. It is untutored in style, and French mapmakers from Louisbourg imprisoned in London would have had little opportunity, or desire, to create a flamboyant record of this devastating defeat. The map is not drawn to scale but provides an primitive bird's-eye view of the siege that effectively destroyed the Canadians' homes and way of life. A detail of the map is shown opposite.

In the 1960s, restoration began on the two-hundred-year-old ruin of Louisbourg. Today, with only one quarter of the fortress reconstructed, it is one of Canada's top tourist sites. Upon passing through the main gate, the first building one encounters is the home of Pierre-Jerome Lartigue.

Detail, figure 14, pp. 30–31

The 1759 Battle of Quebec

Charles Dickens characterized Quebec as "this Gibraltar of America" and in his *American Notes* (1842) he reflected on an epic battle that had taken place there: "The dangerous precipice along whose rocky front, Wolfe and his brave companions climbed to glory; the Plains of Abraham, where he received his mortal wound; the fortress so chivalrously defended by Montcalm; and his soldier's grave, dug for him while yet alive, by the bursting of a shell." The mythical status accorded this final major battle of the French and Indian War overshadows another development that occurred during the three-month siege. A golden era of British military cartography had its beginnings at Quebec.

At the time of the French and Indian War, most of the maps then in circulation recorded military engagements retrospectively. Patrick Mackellar possessed a map that served a different purpose: it was useful in planning an actual battle. Mackellar had been severely wounded serving with General Braddock at the Monongahela in 1755 but was nevertheless able to oversee construction of the fortifications at Oswego the following year. Before much progress had been made, however, the French overwhelmed the garrison and Mackellar was captured and taken to Quebec. In a prisoner exchange nine months later he was sent to London, where he promptly presented the Ordnance Board a detailed description of the Quebec fortifications along with his map. The signed map (figure 15) was so valuable that Mackellar was appointed chief engineer of the Quebec campaign under the command of General James Wolfe. The map shows the fortified town perched atop the steep cliffs (shaded) of Cape Diamond ("The Rock") at the confluence of the St. Lawrence and St. Charles Rivers. After examining it, Wolfe clearly understood why Mackellar believed only an attack against the landward side of the town could be successful.

In May 1759 the British admiral Charles Saunders prepared to transport Wolfe's 8,500 troops and supplies from Louisbourg to Quebec. The 850-mile voyage required navigating the difficult St. Lawrence River. Wolfe had written Amherst the year before that "the aversion to that navigation, and the apprehensions about it, are inconceivably great." Despite the obstacles, the fleet set sail on June 6, 1759, as Wolfe's 8,500 men cheered "British colours on every French fort, port and garrison in North America."

The St. Lawrence flows in a northward direction. The British fleet sailed steadily against the current until the ships neared Quebec where the conditions on the river worsened. The most foreboding obstacle was the Traverse, situated 30 miles below Quebec at Orleans Island, where cliffs rise dramatically and the river itself divides. Even the most experienced French pilots, equipped with charts and marker buoys, dreaded this stretch as it abounds in shallows and jagged rock formations. Convinced that the Traverse was

Figure 15. *Plan of the Town of Quebeck in Canada*, Patrick Mackellar, ca. 1757–58, manuscript, 41 × 36 cm. © The British Library Board (Maps K. Top 119.34)

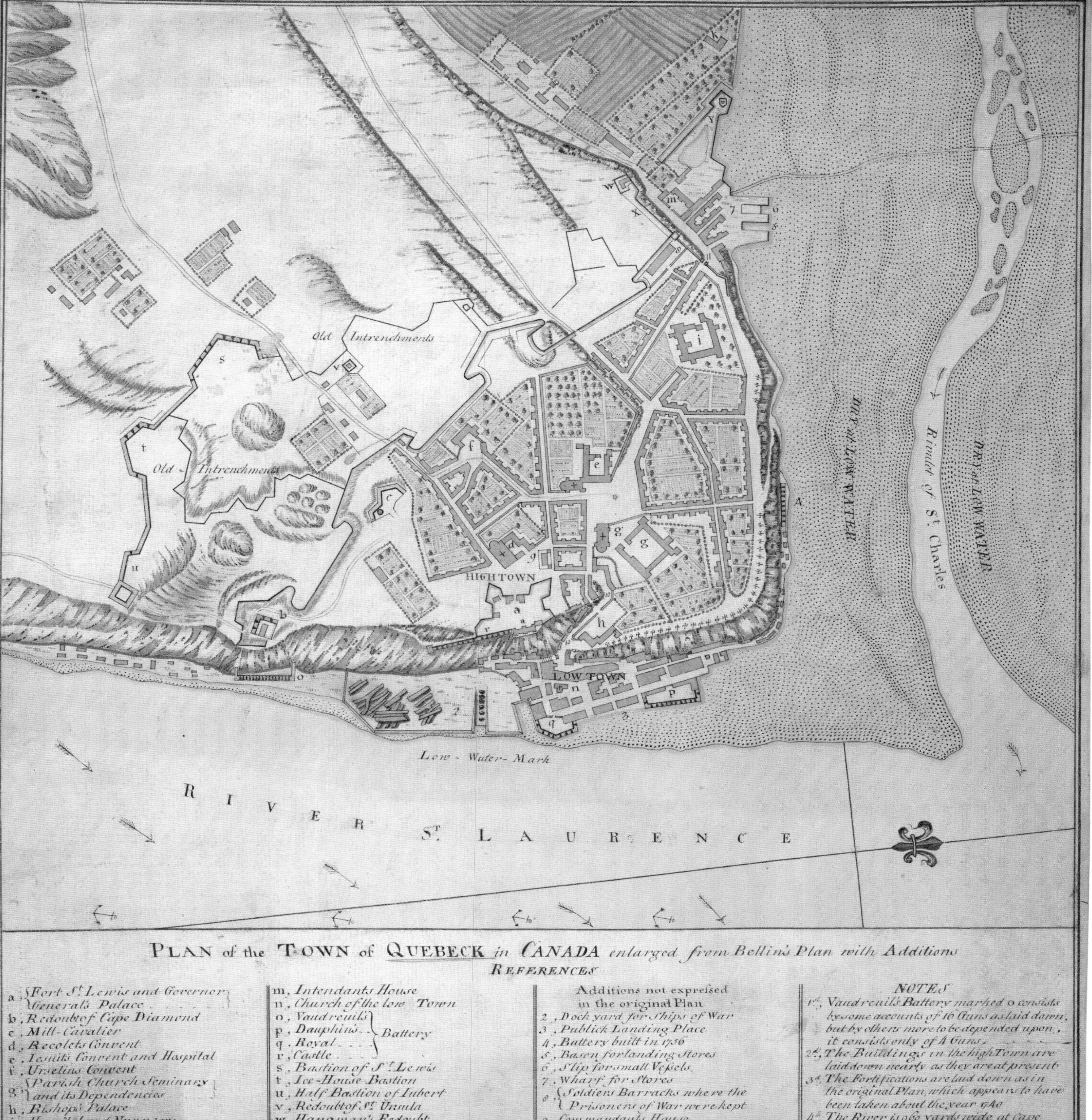
Old Intrenchments
Old Intrenchments
HIGH TOWN
LOW TOWN
Low - Water - Mark
RIVER St. LAURENCE
DRY at LOW WATER
Rivulet of St Charles
DRY at LOW WATER
PLAN of the TOWN of QUEBECK in CANADA enlarged from Bellin's Plan with Additions
REFERENCES
a, Fort St. Lewis and Governor General's Palace
b, Redoubt of Cape Diamond
c, Mill-Cavalier
d, Recolets Convent
e, Iesuits Convent and Hospital
f, Urselins Convent
g, Parish Church Seminary and its Dependencies
h, Bishop's Palace
i, Hospital and Nunnery
k, Church of St. Rock
l, A Battery of 57 Guns without a Parapet
m, Intendants House
n, Church of the low Town
o, Vaudreuil's
p, Dauphin's
q, Royal
r, Castle
Battery
s, Bastion of St. Lewis
t, Ice-House-Bastion
u, Half Bastion of Iubert
v, Redoubt of St. Ursula
w, Hangman's Redoubt
x, Gallow's Hill Line
y, Redoubt of St. Rock
Additions not expressed in the original Plan
2, Dock yard for Ships of War
3, Publick Landing Place
4, Battery built in 1756
5, Bason for landing Stores
6, Slip for small Vessels
7, Wharf for Stores
8, Soldiers Barracks where the Prisoners of War were kept
9, Commandants House
10, Principal Communication between ye. two Towns on the East Side
11, Principal Communication on ye. N.W. Side
NOTES
1st, Vaudreuil's Battery marked o consists by some accounts of 16 Guns as laid down, but by others more to be depended upon, it consists only of 4 Guns.
2d, The Buildings in the high Town are laid down nearly as they are at present
3d, The Fortifications are laid down as in the original Plan, which appears to have been taken about the year 1740
4th, The River is 960 yards wide at Cape Diamond & 1130 at the royal Battery.
5th, Common spring Tides rise 18 feet, and equinoctial Springs 25 feet.
SCALE of 400 Feet to an Inch
200 0 400 800 1200 1600 2000
CX/X
34

impassable, the French had not bothered to fortify Cap Tourmente, the commanding position 1,800 feet above the river. This proved to be a costly mistake as Wolfe was able to navigate these waters with relative ease with the aid of a sea chart drawn by James Cook.

Cook, the master, or navigator, of *The Pembroke,* had refined his chart-making skills under Major Samuel Holland, an accomplished surveyor and one of the engineers closest to Wolfe. Cook and Holland had sailed ahead of the fleet, charting the entire St. Lawrence below Quebec, working at night to avoid French guns. The signed manuscript chart (figure 16) was the most important that Cook produced, though in time he would become a world-renowned cartographer, explorer, and circumnavigator. Cook's chart provided a guide for the heavily laden fleet to move gingerly across the Traverse on June 20 without losing a single ship. The next morning, a stunned Canadian governor, Pierre-François de Vaudreuil, wrote the French minister of marine and colonies that "The enemy . . . has passed sixty ships of war where we hardly dared risk a vessel." He hastened to add, however, that this was of little consequence as the fortifications at Quebec were "impenetrable."

The commanding French general, the Marquis de Montcalm, had defeated the British at Ticonderoga in 1758, but he later abandoned the fortress to lead the defense of Quebec. The combined force of three thousand French regulars and fourteen thousand Canadian troops was considerably larger than Wolfe's, but Montcalm doubted that the Canadians would hold their lines in open battle. He deployed his superior numbers in guarding the riverbanks between the St. Charles River and Montmorency Falls several miles below the town. Montcalm intended to wait Wolfe out. By summer's end the British would have to set sail to avoid being iced in.

On June 27 Wolfe disembarked his army and artillery at Pointe Lévy, a mile across the river from Quebec, and attempted to draw out the French forces by reducing the town to rubble and laying waste to nearby farms and villages. The cannonading continued throughout July, inflicting considerable damage to the fortress but producing no signs of surrender. Wolfe concluded a direct attack was necessary. On July 31 a sizable force was sent downriver to gain a beachhead near Montmorency Falls. The French were able to repel the attack despite the spectacular storm that raged during the battle. Upriver, Brigadier General James Murray met with no more success. He had safely moved six ships past Quebec's cannon, but Montcalm dispatched some three thousand troops under Louis-Antoine de Bougainville, and they easily thwarted Murray's attempts to land. A triumphant Vaudreuil wrote to General François-Charles Bourlamaque, "I have no more anxiety about Quebec. M. Wolfe, I can assure you, will make no progress."

The health of chronically ill Wolfe took a bad turn in mid-August. From his sickbed he planned a final attack, but on August 30 he notified Admiral Saunders that "My ill-state of health hinders me from executing my own plan; it is of too desperate a nature to order others to execute."

Gloom and melancholy permeated the British camp. Brigadier General George Townshend, in a September 6, 1759, letter to his wife, foresaw an early return home. "Gen. Wolf's health is but very bad. His generalship in my poor opinion–is not a bit better." But Wolfe rallied. "I am so far recovered as to be able to do business," he wrote to the Earl of Holderness, "but my constitution is entirely ruined, without the consolation of having done any considerable service to the state, or with-out any prospect of it."

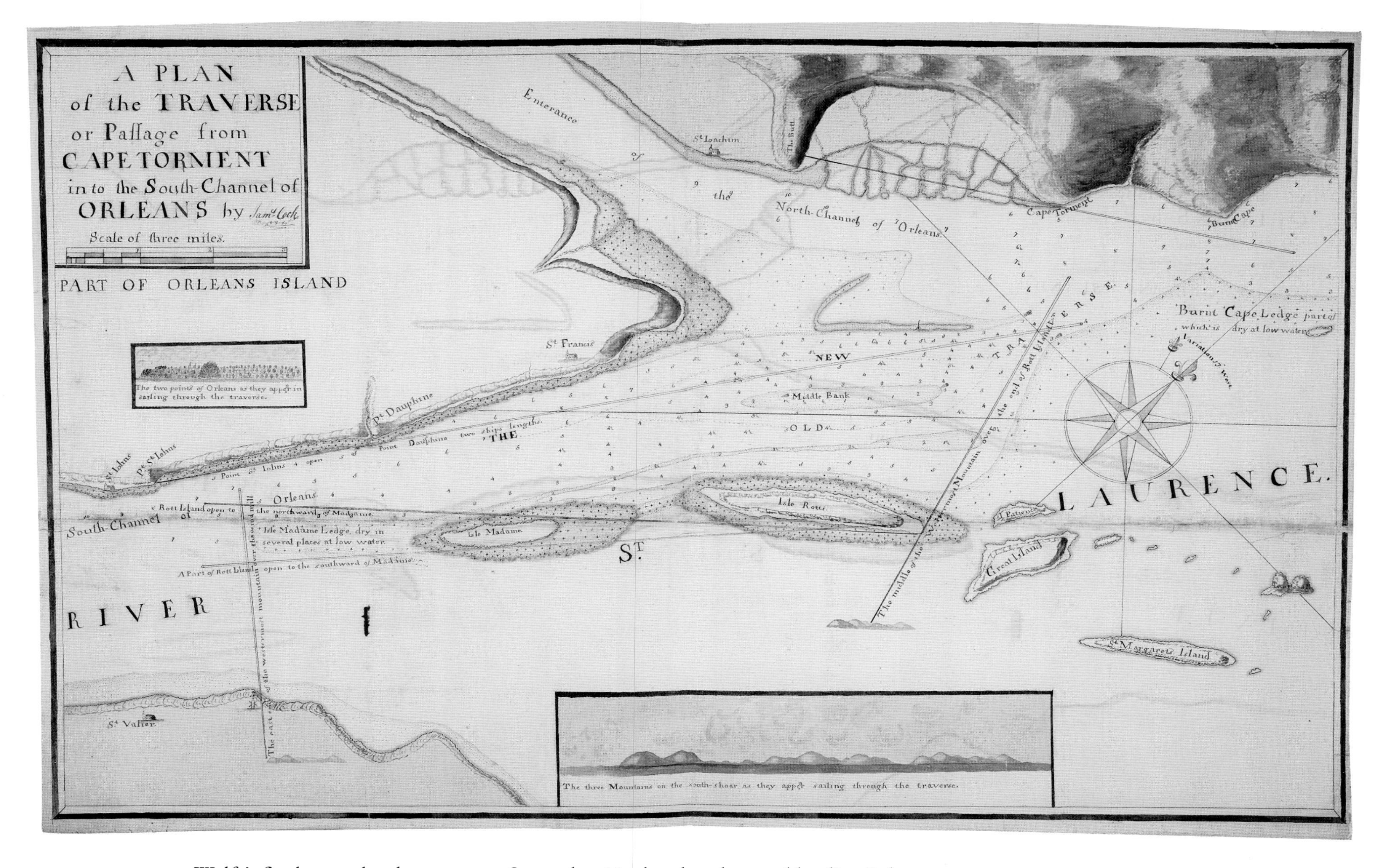

Figure 16. *A Plan of the Traverse or Passage from Cape Torment*, James Cook, ca. 1758–59, manuscript, 51 × 79 cm. © The British Library Board (ADD. 31360 f)

Wolfe's final general orders came on September 12 when he advocated landing "where the French least expect it." The choice of the Anse-au-Foulon, a small cove three miles upriver at the base of a 300-foot-high cliff, was as shocking to Wolfe's brigadiers as it was to the French. In an extraordinary feat of seamanship, the navy disembarked the army at night. Many soldiers doubted that the scrub-covered cliff could be scaled but an advanced guard of light grenadiers under Captain William Howe reached the peak and overwhelmed a garrison of Canadian militia. Sixteen years later the same William Howe would order and lead the British assault on Bunker Hill.

Snaking an army up a cliff and placing it between the forces of Montcalm and Bougainville was so audacious that it was almost suicidal. It meant that the outcome would be either a quick victory or total defeat. The navy had no plans to facilitate a retreat and the troops carried no entrenching tools.

When Montcalm awoke on September 13, Wolfe's army was nowhere to be seen. The camp on the Isle d'Orleáns was abandoned and the British troops were assembling on the Plains of Abraham. Montcalm ordered a preemptive strike without waiting for support from Bougainville. After three months of cautious maneuvering, the only European-style battle of the French and Indian War was decided within thirty minutes. "When they had arrived within 100 yards of our line," Mackellar wrote, "our troops advanced regularly with

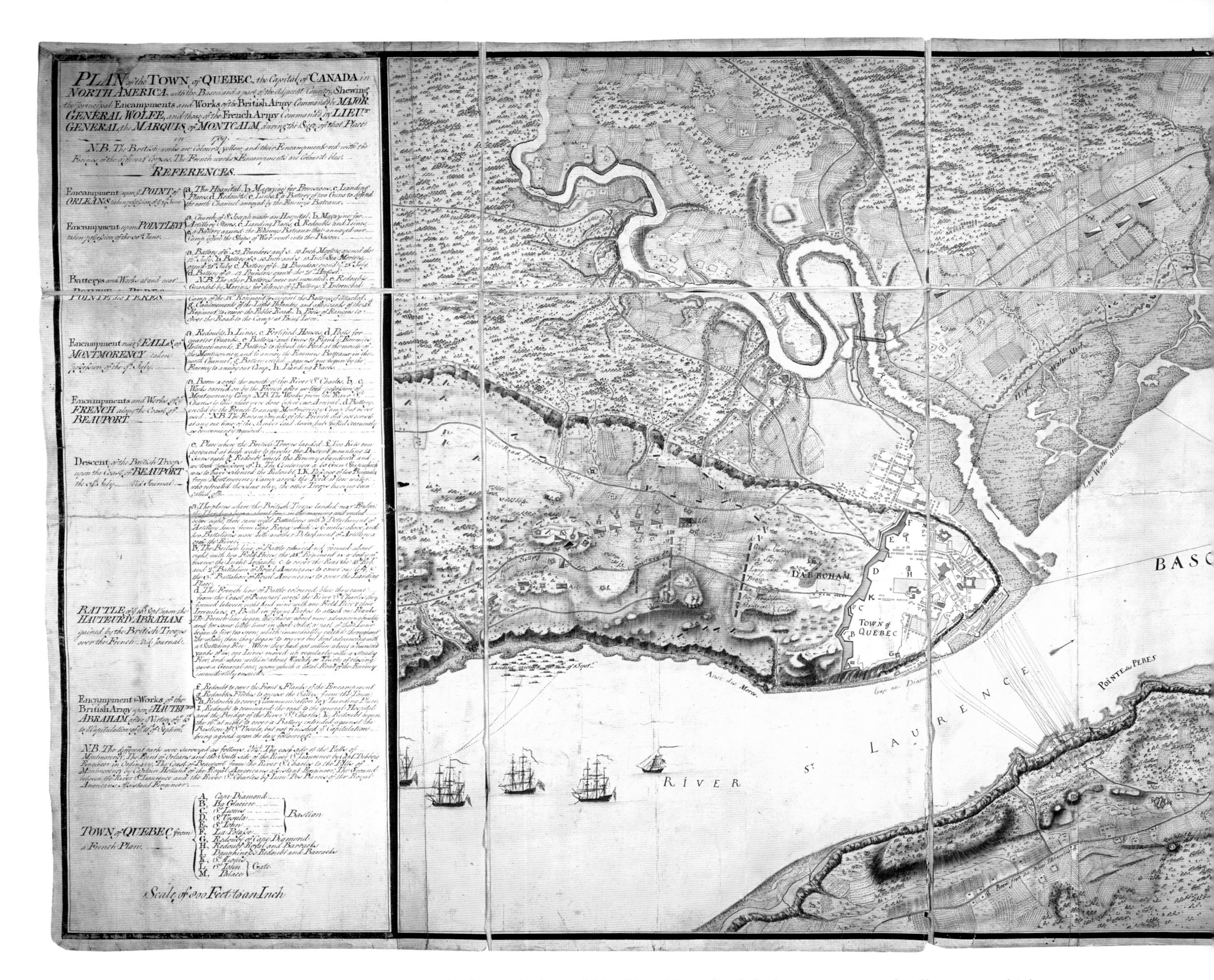

Figure 17. *Plan of the Town of Quebec, the Capital of Canada*, Captn. Debbeig/Captain Holland/& Lieut. Des Barres, 1759, manuscript, 71 × 181 cm. © The British Library Board (ADD 31357 A)

a steady fire, and when within 20 or 30 yards of closing gave a general volley, upon which the enemy's whole line turned in the same instant, and fled in a most precipitate manner." Wolfe appeared on the battlefield in his colorful general's uniform and was shot four times by snipers. Montcalm, mounted upon his horse, received his mortal wound at almost the same moment that Wolfe received his fatal gunshot.

Fortune often plays a defining role in battles that change the course of history and so it did during the British victory at Quebec. As Governor Vaudreuil observed, "At no other hour and no other spot of ground would the French army be defeated." There would be several more years of conflict between Britain and France but the outcome had been

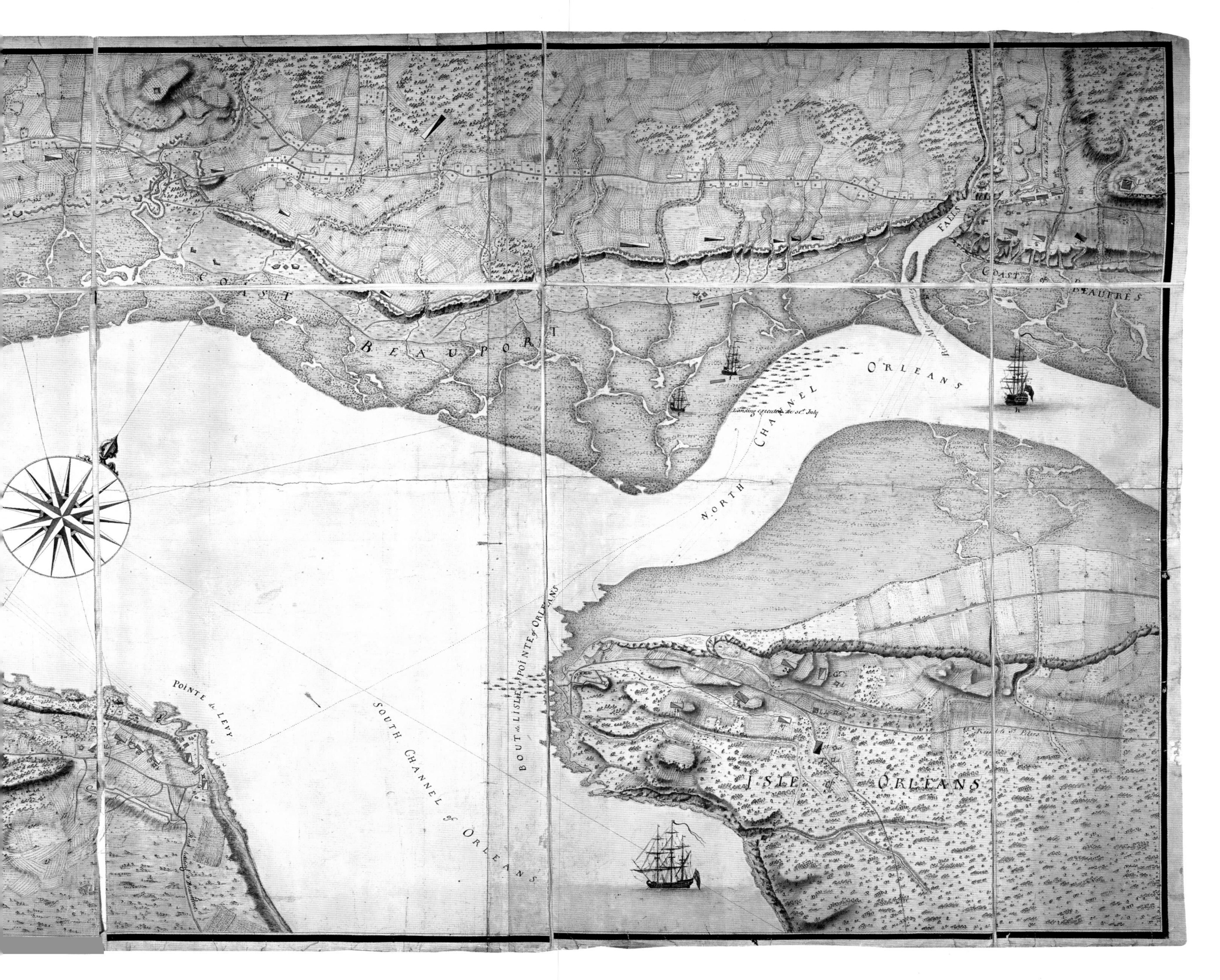

determined. In 1763 the Treaty of Paris awarded the British an empire in North America. Twenty years later, much of that empire would be relinquished following an American victory at Yorktown. During the interim, cartographers present at Quebec would extensively map British possessions in North America and record the great events of the American Revolution. Two of the most important, Samuel Holland and J.F.W. Des Barres, are credited with this exemplary map of the entire Quebec theater of action (figure 17). Its sophistication is striking when compared to the primitive map that Mackellar had provided to Wolfe at the onset of the battle.

pages 40–41: Detail, figure 21. *View of Boston, the Capital of New England*, William Pierie, 1773, p. 46

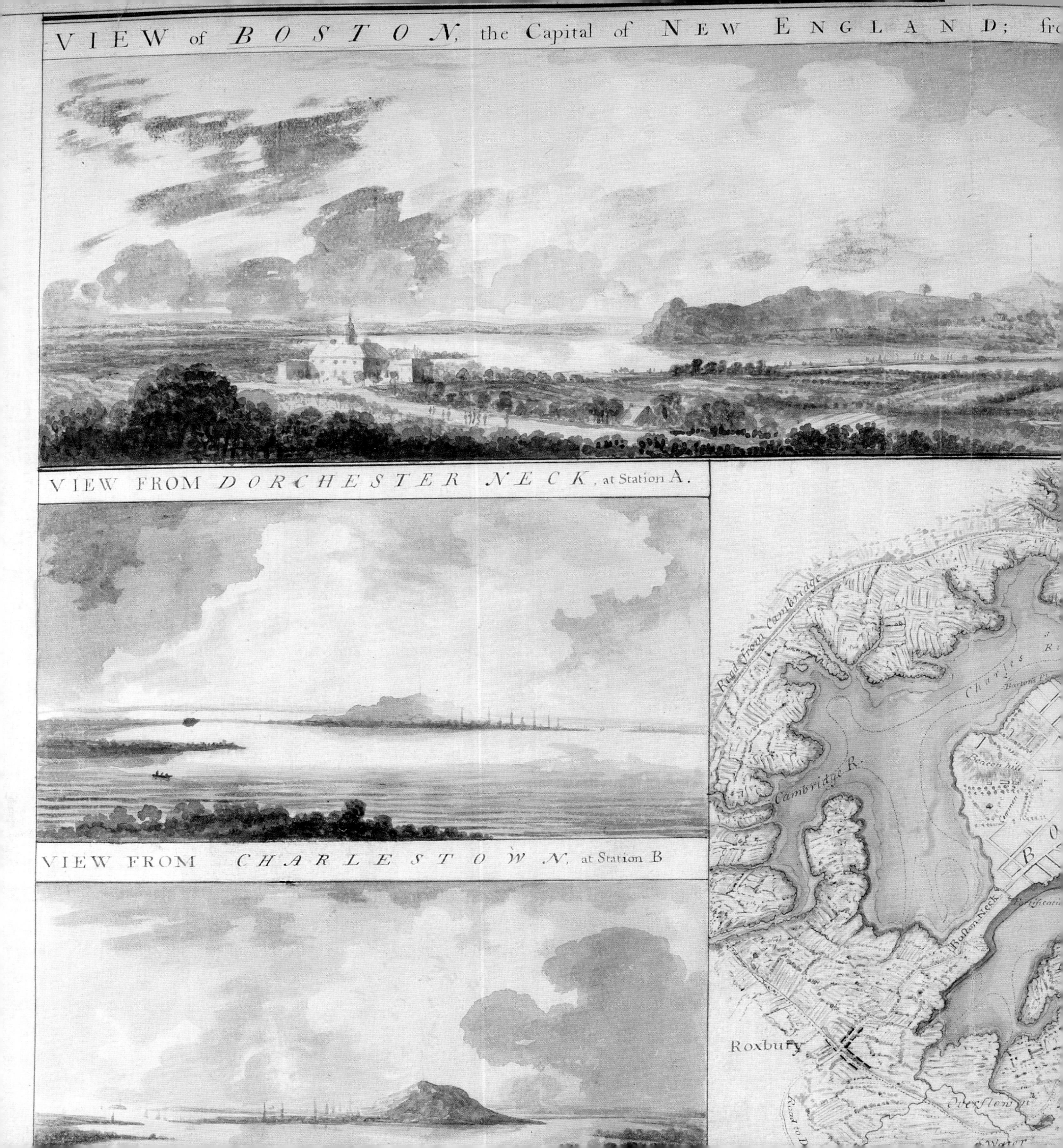
VIEW of BOSTON, the Capital of NEW ENGLAND; fr
VIEW FROM DORCHESTER NECK, at Station A.
VIEW FROM CHARLESTOWN, at Station B
Road from Cambridge
Charles R
Cambridge R.
Beacon hill
Common
Boston Neck
Roxbury
Overflown
B

Between the Wars

British Military Engineers Map North America

The Death of General Wolfe (figure 18) is one of the most iconic images of colonial North America. Painted in 1770 by Benjamin West, it glorifies James Wolfe's heroic sacrifice during the British victory at Quebec. West's finest historical painting was immediately acclaimed throughout the British empire. It is a visual powerhouse, but the scene portrayed is not reality. Only Lieutenant Henry Browne, seen holding the flag, was present at Wolfe's death. The other officers and the symbolic "noble savage" are notables that West felt would be fitting observers at Wolfe's demise. Engineer Samuel Holland, for one, rued the fact that he had been "attendant on the general in glorious exit, but others are exhibited in that painting who never were in battle." Neoclassical painting provided an unrealistic depiction of the people and landscape of North America.

The most accurate visual record of North America between the French and Indian War and Revolutionary War survives in the artistic endeavors of British engineers and artillery officers. Many of these talented officers, including those featured here, were graduates of the Royal Military Academy at Woolwich where "gentleman cadets" were taught surveying, architectural drawing, and fortification planning. Cadets were also expected to become accomplished topographers who could produce "accurate descriptions of particular places." The Woolwich engineer used his kit containing pen, ink, and watercolors to faithfully record these places in both manuscript maps and landscape views.

Paul Sandby, a man who never set foot in North America, was the standard bearer for Woolwich-trained topographers. In 1749, at age twenty-one, Sandby produced a monumental map of Scotland that Dr. Yolande Hodson, an expert in British royal collections, describes as "an exceptional work of art . . . one of the jewels in the Royal Topographical collections and Britain's national heritage." Sandby was also a remarkable watercolorist. He is credited with founding the English School of Watercolor and introducing advancements to a process he named "aquatint," a procedure ideally suited to printing watercolor images. The plans and views shown in figure 19 provide an early glimpse of his skills as both

Figure 18. *The Death of General Wolfe*, Benjamin West, 1770, oil on canvas, 153 × 215 cm. National Gallery of Canada, Ottawa

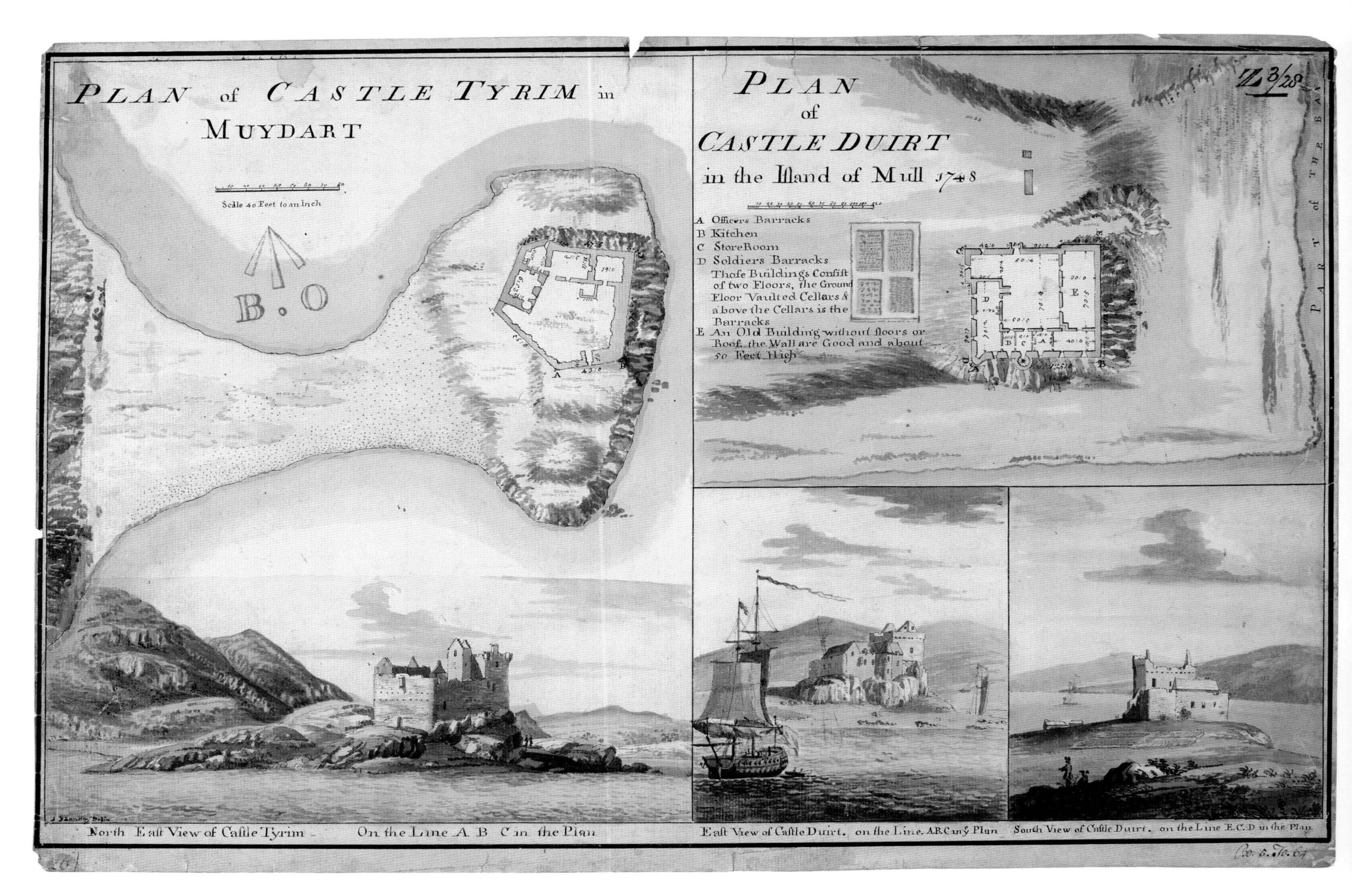

Figure 19. *Plan of Castle Tyrim*, Paul Sandby, 1748, manuscript, 36 × 55 cm. © National Library of Scotland

mapmaker and landscape watercolorist. Years later, Thomas Gainsborough would credit Sandby as "the only man of genius," who painted "*real Views* from nature in this Country." In 1768 Sandby became a founding member of the Royal Academy and he was appointed Drawing Master at Woolwich, a position that he would hold for thirty-one years.

No one produced more maps, sea charts, and views in North America than Joseph Frederick Wallet Des Barres. A talented Protestant Huguenot, Des Barres left France to attend Woolwich in 1753, under the patronage of the Duke of Cumberland, the leading British military figure of the time. Des Barres contributed to the spectacular map of Quebec shown on pages 38–39. In 1763 he was assigned by the Admiralty to make accurate surveys of the Canadian coasts and harbors. Eventually, Des Barres's surveys would cover the entire Atlantic coast, encompassing over 15,000 miles of exceptionally complex and difficult shoreline. The first editions of the charts were rushed into print in 1775 and bound into the four-volume *Atlantic Neptune* atlas described by the esteemed nineteenth-century bibliophile Obadiah Rich as "the most splendid collection of charts, plans and views ever published."

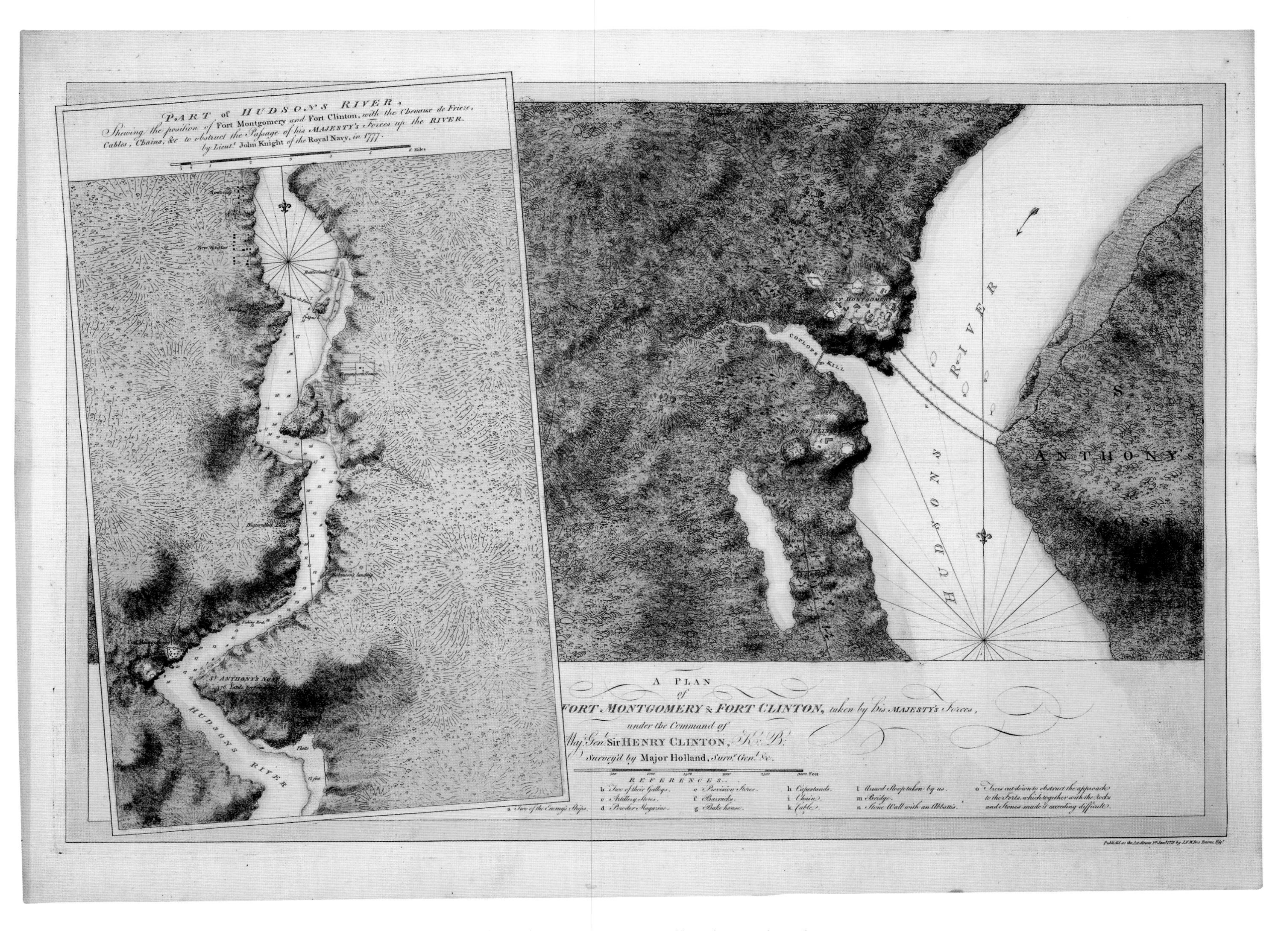

Figure 20. *A Plan of Fort Montgomery & Fort Clinton*, J. F. W. Des Barres/Samuel Holland, London, 1779, Copperplate engraving, 55 × 80 cm. © National Maritime Museum, Greenwich, London

Many talented Woolwich officers contributed to Des Barres's effort but only a few of the mapmakers are identified. Samuel Holland was credited for the striking map of Fort Montgomery and Fort Clinton (figure 20). Holland had produced the manuscript source for this map now in the William Faden Collection at the Library of Congress. The *Atlantic Neptune* contains more than 140 views. Of these, Stephanie Cyr of the Norman Leventhal Map Center has found only the work of artillery officer William Pierie directly credited. In 1773 Pierie produced the exquisite manuscript map surrounded by six views of Boston shown in figure 21. All of these views appear in various formats in the *Atlantic Neptune*.

In 1766 Bernard Ratzer began work on the finest printed map of an American city produced in the eighteenth century. *This Plan of the City of New York and its Environs* shows cultivated fields, forests, and salt meadows, interspersed with large estates, dominating the modest city of twenty thousand at Manhattan's southern tip. The first edition of the "Ratzer Map" was published in 1770 and is known in only three copies. The example in figure 22 is from George III's Topographical Collection and its spectacular

Figure 21. *View of Boston, the Capital of New England*, William Pierie, 1773, manuscript, 56 × 89 cm. © The British Library Board (maps K. Top 120.34)

Figure 22. *Plan of the City of New York in North America: Surveyed in the Years 1766 & 1767*, Bernard Ratzer, London, ca. 1770, copperplate engraving, 118 × 86 cm. © The British Library Board (K. Top 121.36b)

contemporary coloring suggests that it was a presentation copy given to the king by the publisher Thomas Kitchen. In 1776, Faden and Jeffreys published a more widely known second edition, not illustrated here, in response to the heightened interest in the American conflict.

Artillery Lieutenant Thomas Davies drew the view from Governors Island at the bottom of the Ratzer map. It shows Manhattan from the Battery to the location of the present-day Williamsburg Bridge. Davies produced a number of watercolor paintings of North America during the 1760s and is described by R. H. Hubbard as "the most talented of all the early topographical painters in Canada." In 1776 Davies would assume command of the artillery at Fort Knyphausen after it was captured from the Americans. Several of Davies's drawings survive as a rare visual record of that battle.

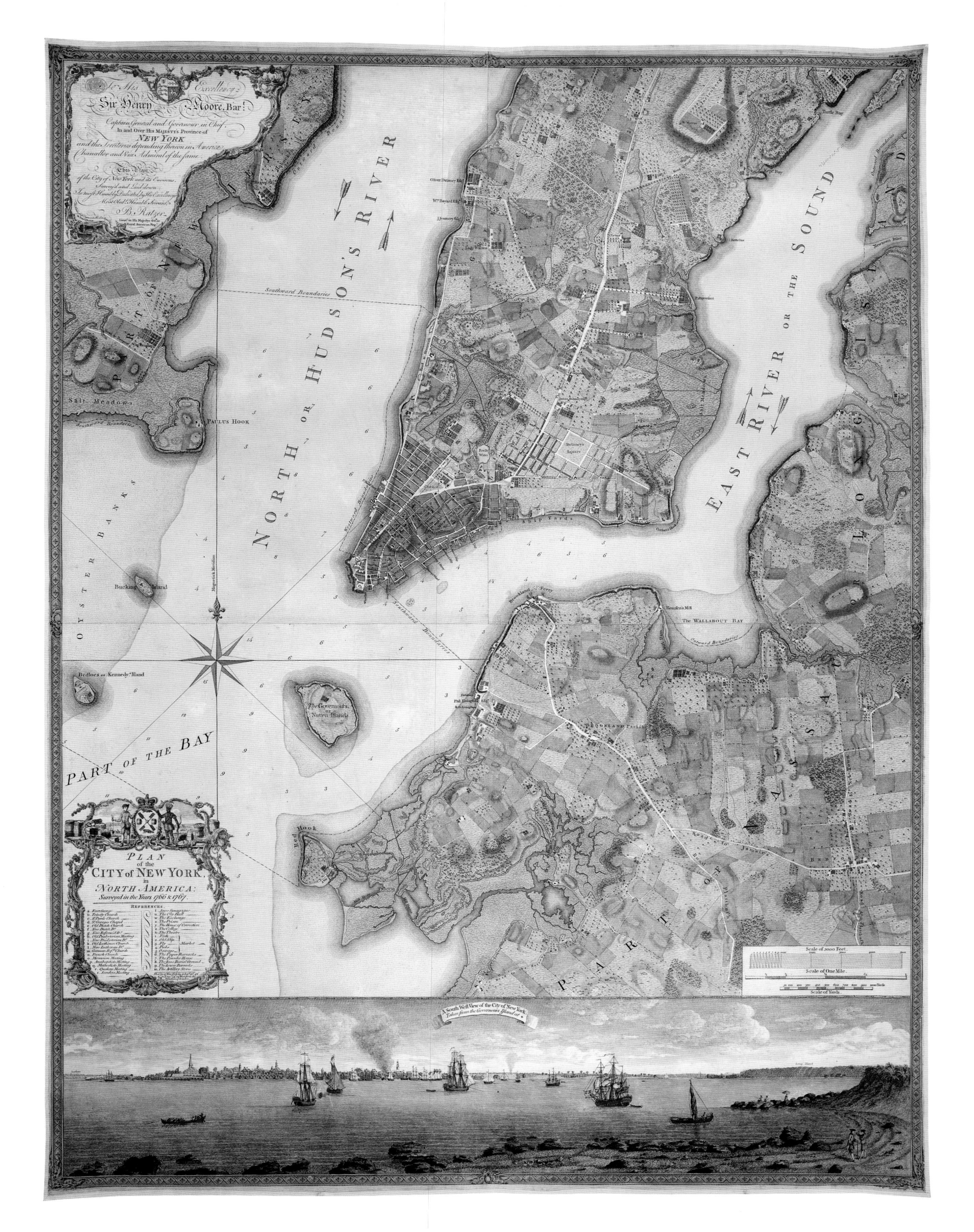
To His Excellency
Sir Henry Moore, Bar.t
Captain General and Governour in Chief
In and Over His Majesty's Province of
NEW YORK
and the Territories depending thereon in America
Chancellor and Vice Admiral of the same
This Plan
of the City of New York and its Environs
Survey'd and Laid down
Is most Humbly Dedicated By His Excellency's
Most Obed.t Humble Servant
B. Ratzer
PART OF NEW JERSEY
NORTH OR HUDSON'S RIVER
EAST RIVER OR THE SOUND
Southward Boundaries
Salt Meadows
Paulus Hook
OYSTER BANKS
Bucking Island
Magnetick Meridian
Bedloes or Kennedy's Island
The Governours or Nutten Island
PART OF THE BAY
Delancey's Square
Ship Yards
Brookland Ferry
The Wallabout Bay
Red Hook
PLAN
of the
CITY of NEW YORK,
in
NORTH AMERICA:
Surveyed in the Years 1766 & 1767.
REFERENCES.
Scale of 2000 Feet
Scale of One Mile
Scale of Yards
A South West View of the City of New York
Taken from the Governours Island at

Figure 23. *View from Beacon Hill of North Boston with Noodle Island in the background and the Mill Pond in the left foreground*, Richard Williams, 1775, manuscript, 17 × 48 cm. Private collection

On June 11, 1775, a talented twenty-five-year-old officer in the Royal Welch Fusiliers arrived in Boston on the cusp of revolution. Over the next three months Lieutenant Richard Williams would map and sketch the town and maintain a journal, subsequently published under the title *Discord and Civil Wars.* The entry for June 12 reads: "I went to the common & to Beacon hill, where I saw all our encampments, & those of the Enemy. From this hill you have a view of the town and country all around it." Williams captured these views in a set of six watercolors providing a 360-degree perspective from Beacon Hill. They constitute the most realistic visual legacy of Boston at the onset of the American Revolution. One of the views of Noodle Island and the Mill Pond (long since filled in to create the Back Bay area) is shown in figure 23. The image comes from a recently discovered set of watercolors that, accompanied by pencil sketches, maps, and other drawings, constitutes the artist's complete original portfolio.

On June 17, Williams's unit was held in reserve at Bunker Hill, but he observed the entire battle and incorporated the details into a manuscript map of Boston. On July 28, Williams sent copies of his work to the Ordnance Office in London: "Col: James sailed for

England, I sent letters by him & gave him several of my drawings of the views round Boston &c." The drawings Williams refers to are annotated copies similar to the example illustrated here and kept at the British Library. Williams's manuscript map is also in the British Library and is credited as the source for *A Plan of Boston and its Environs*, published in London by Andrew Dury in March 1776 (figure 24). It is among the finest printed maps of Boston showing this critical period of British occupation.

In August 1775 Lieutenant Williams was directed to lead fifty men onto the *Empress of Russia* and proceed to Nova Scotia. On September 4 his ship "came to Anchor near the Town of Annapolis." Williams began to describe the town but his journal abruptly ends mid-sentence. No subsequent journal is known although a charming group of his paintings of Annapolis Royal resides at the Art Gallery of Nova Scotia. The last we hear of Williams is in an obituary in the *Morning Post & Daily Advertiser.* The London newspaper reported that the lieutenant had "lately come from America for the recovery of his health," but had died at Penryn, in Cornwall, on April 30, 1776. In less than a year he had made a spectacular contribution to the topographical history of North America.

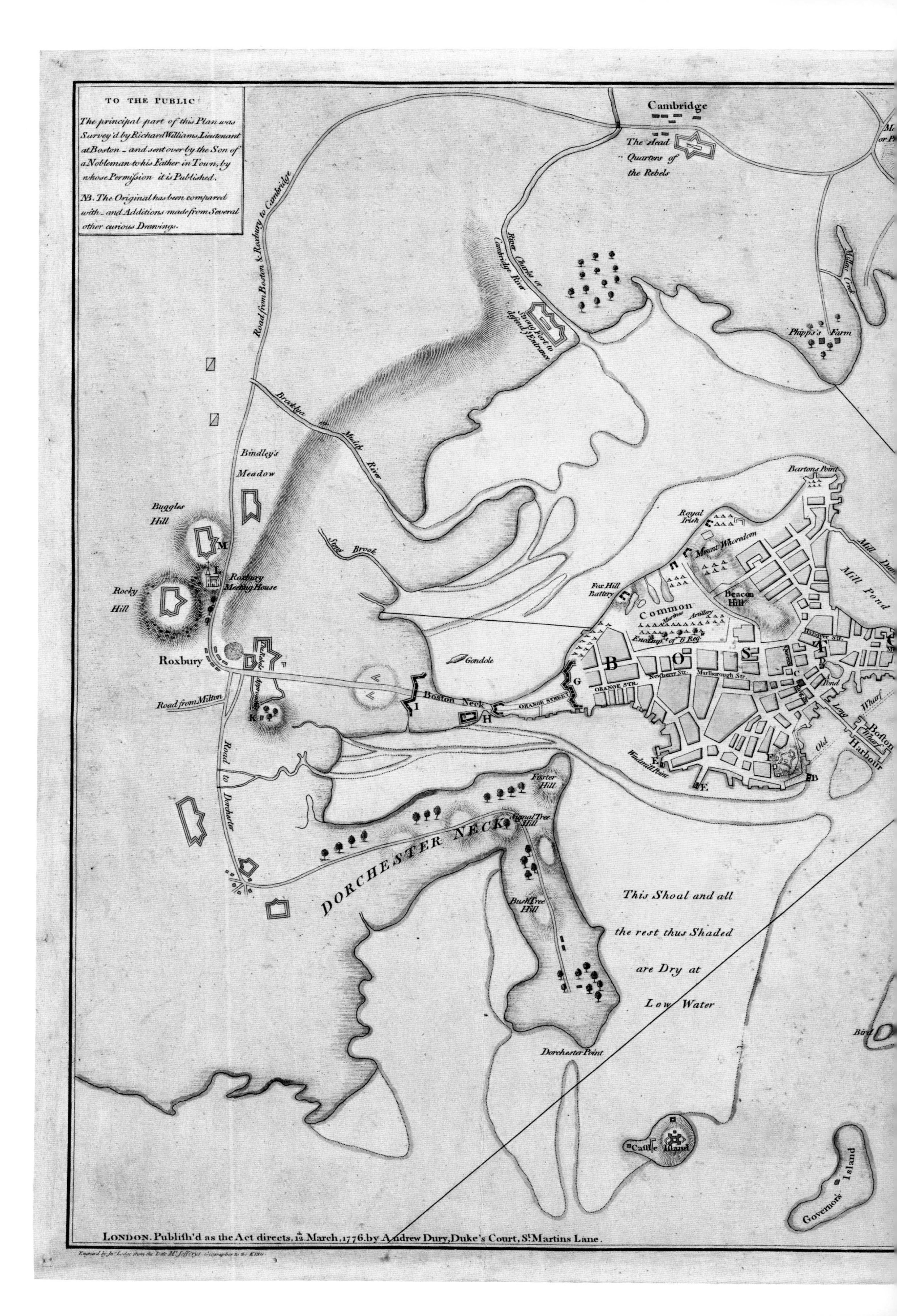

Figure 24. *A Plan of Boston and its Environs*, Richard Williams, London, 1776, copperplate engraving, 45 × 65 cm. Courtesy Mapping Boston Foundation

A PLAN
OF
BOSTON,
and its ENVIRONS.
shewing the true SITUATION of
HIS MAJESTY'S ARMY.
AND ALSO THOSE OF THE
REBELS.
Drawn by an Engineer at Boston. Octr. 1775.
REFERENCE.
A { Corpse Hill, a Battery of 8 Pieces of Cannon, Mortars, &c. &c. Erected to favour the Troops Landing on ye 17th June & to set Fire to Charles Town
B { North & South Batteries, built by the Province for the defence of ye Harbour, they are in a ruinous State
C. Town Hall
D. Faneuil Hall
E. Two Batteries Erected on Wharfs against Dorchester Neck
F. Fort Hill, a proper Place for constructing a Citadel
G. Fortification now constructing for the immediate defence of the Town
H. A Block House & 2 Strong Batteries pointing on part of Dorchester Neck
I. Lines to defend the Boston Neck
K { A Hill from whence the Enemy annoy ye Centries & Officers with small Arms, but seldom do any Execution
L { Roxbury Meeting House, upon a Hill from whence ye Enemy often Fire Cannon into the Lines
M { A Strong Post of the Enemy Fortified in appearance with great Judgment, & much Elevated, from whence with a 24 Pounder they can just reach the Lines
EXPLANATION.
The Works shaded Green shew those constructed by His Majesty's Troops
The Works shaded Yellow shew those thrown up by the Rebels, as they appear from Boston.
Plough'd Hill
Winter Hill
Mystick River
Penny Ferry
Charles Town Neck
Lines & Redoubts thrown up by our Troops after ye Victory on ye 17 June 1775.
Gondoles
Artillery
Gen! Howe's Camp
HILL
Dragoons
Pond
Marines
Charlestown Point
Troops Landed 17 June under General Howe
North End
Road to Marble Head & Salem
Winnisimmet Ferry
NODDLES ISLAND
Williams's House burnt by ye Rebels.
HOG ISLAND
Yards or Half a Mile

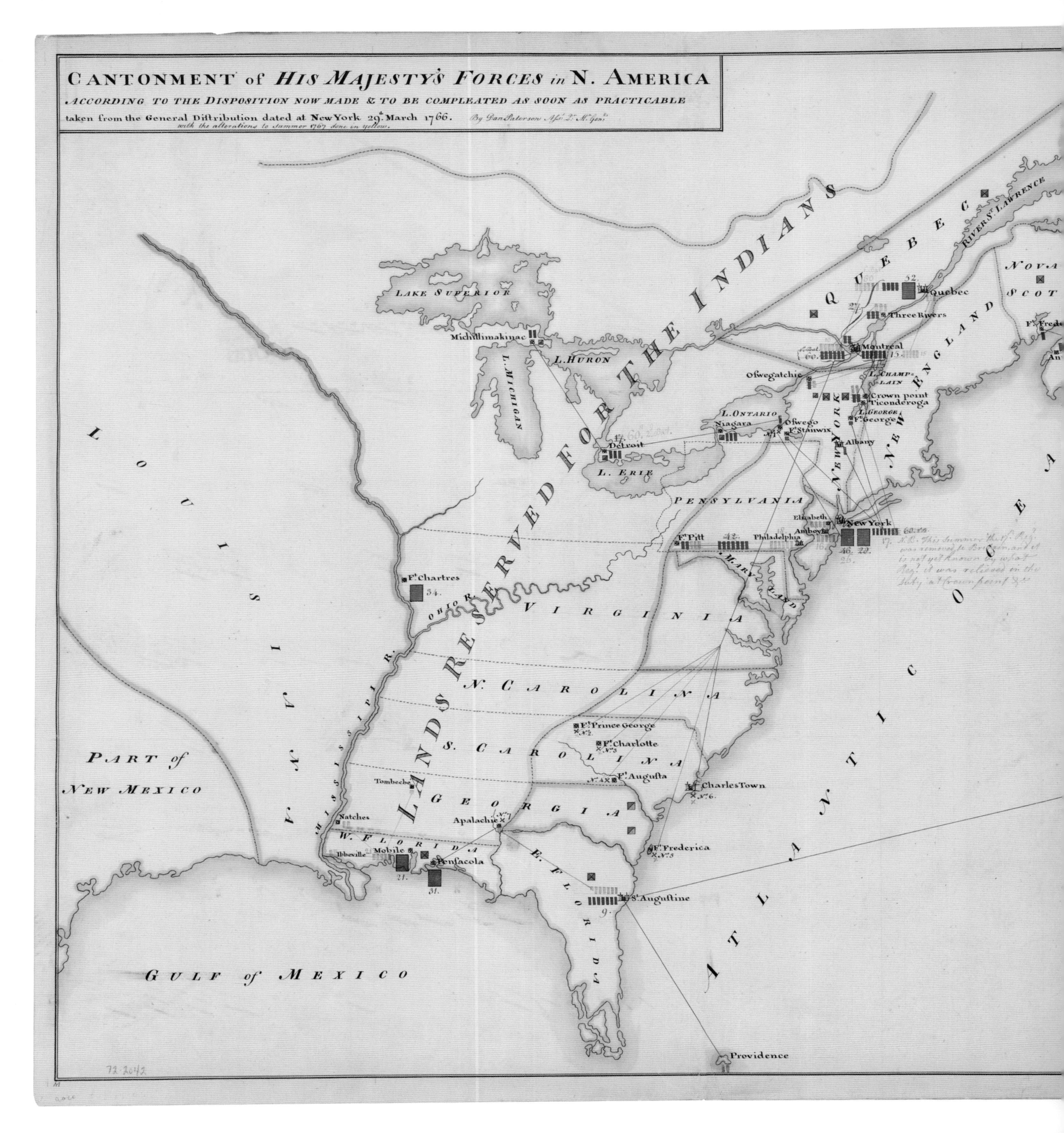

CANTONMENT of HIS MAJESTY'S FORCES in N. AMERICA
ACCORDING TO THE DISPOSITION NOW MADE & TO BE COMPLEATED AS SOON AS PRACTICABLE
taken from the General Distribution dated at New York 29th March 1766. By Dan Paterson Ass. Qr. Mr. Genl.
with the alterations to summer 1767 done in yellow.
LANDS RESERVED FOR THE INDIANS
QUEBEC
RIVER St. LAWRENCE
NEW ENGLAND
NEW YORK
PENSYLVANIA
MARYLAND
VIRGINIA
N. CAROLINA
S. CAROLINA
GEORGIA
W. FLORIDA
E. FLORIDA
LOUISIANA
PART of NEW MEXICO
GULF of MEXICO
ATLANTIC
Lake Superior
L. Michigan
L. Huron
L. Ontario
L. Erie
L. Champlain
L. George
Ohio R.
Missisippi R.
Quebec
Three Rivers
Montreal
Oswegatchie
Crown point
Ticonderoga
Ft. George
Albany
Niagara
Oswego
Ft. Stanwix
Detroit
Michillimakinac
New York
Elizabeth
Amboy
Philadelphia
Ft. Pitt
Ft. Chartres
Ft. Prince George
Ft. Charlotte
Ft. Augusta
Charles Town
Ft. Frederica
St. Augustine
Apalachie
Tombeche
Natches
Ibbeville
Mobile
Pensacola
Providence

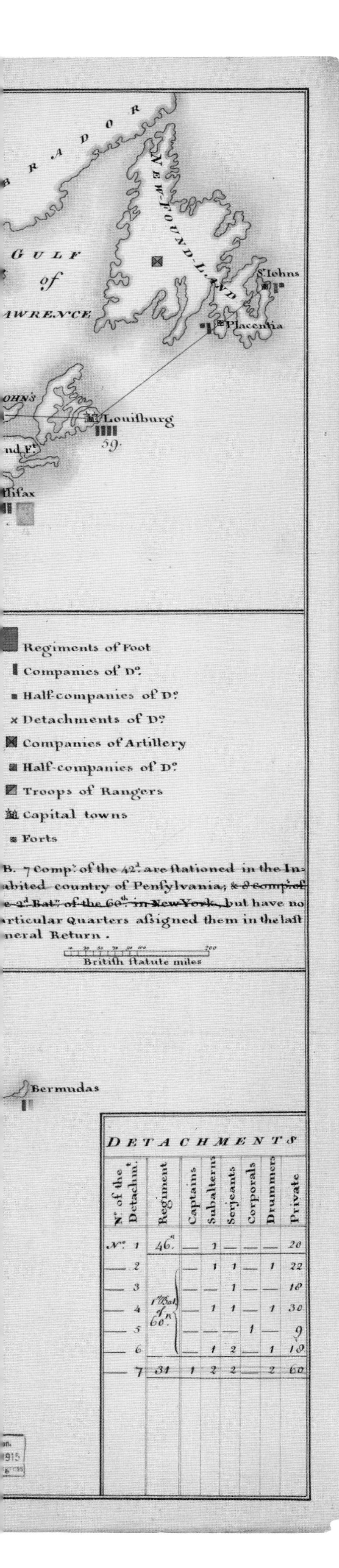

The Cantonment Maps

More territory changed hands when the Treaty of Paris was signed on February 10, 1763, than at any time before or since. On that day, England gained clear title to most of North America. Over the coming years the British followed the peacetime pattern of cantoning (encamping) forces along the periphery of the colonies in Quebec, Halifax, and Mobile, with a few troops scattered among outposts on the continent's interior. By 1765 encroachments from the Indians, the Spanish, and the French heightened tensions, and the British were compelled to respond to these threats. But as Fred Anderson has pointed out in *Crucible of War*, "However effective the redcoats had once been as conquerors, they were utterly unsuited to controlling the conquests."

The British army had limited troops at the ready in America but took advantage of these small numbers by strategically moving them from place to place. Within two and a half years of the peace fifteen regiments, a total of some six thousand troops were stationed in various sites in North America. The cantonments spread from the coastlines to forts along rivers and lakes; they had served as centers of communication and trading during the first half of the eighteenth century. As British troops crisscrossed the country in 1765–67 they rebuilt forts, abandoned others, and marched into areas of conflict. These comings and goings are meticulously recorded in *The Correspondence of General Thomas Gage, 1763–1775*. As commander of military forces in North America, General Gage's diligent communications deliver an almost day-by-day accounting of the exigencies of army life.

At the same time that Gage was ordering troops from place to place, a series of revealing manuscript maps were drafted to track the exact locations of these troops, along with the numbers of their regiments and size. One is titled *Cantonment of His Majesty's Forces in N. America* and the others all have similar titles. The earliest is dated October 11, 1765, and is located at the Library of Congress. A second specimen at the Clements Library, University of Michigan, is dated simply 1766, though it appears to have been drafted very early in 1766 because it precedes the

Figure 25. *Cantonment of His Majesty's Forces in N. America . . . taken from the general distribution dated . . . 29th March 1766 with the alterations to summer 1767 done in yellow*, Daniel Paterson, 1767, manuscript, 51 × 61 cm. Geography and Map Division, Library of Congress

Figure 26. *General Thomas Gage*, John Singleton Copley, 1768, oil on canvas, 127 × 101 cm. Yale Center for British Art, Paul Mellon Collection

important changes in troop dispositions that are indicated on the British Library copy dated March 29, 1766. The last known map with "alterations to the summer 1767" is at the Library of Congress and is illustrated in figure 25. All of these maps seem to be drawn from many of the same records that Gage (figure 26) had at his disposal and are a visual transcript of the part of his correspondence that relates to British troop movements in America.

The probable mapmaker of the entire series was Ensign Daniel Paterson, who signed the third and fourth maps in the series; the first two are unsigned. He was commissioned in 1765 and served with the Quartermaster Corps in London. This branch of the army was responsible for the essentials of warfare: road building, surveying, troop movement, and supplies. When forces were ordered to be distributed throughout North America, Paterson was assigned to map the cantonments. Created in London, the remarkable series of modern, thematic maps of North America came from Paterson's drafting table in the recently completed New Horse Guards building. This was one of the most famous and attractive military buildings in the world and the Italian landscape painter Canaletto immediately painted it, as he had the decaying Old Horse Guards.

In Paterson, the British had selected a man with myriad talents. In addition to his skills as a topographical engineer, he was a formidable statistician. So nimble was he with numbers that he later published a dictionary of 50,000 distances between the principal cities in Great Britain. "Paterson's Roads" became so indispensable to the army that no military exercise within the country was planned without first consulting his intricate calculations.

Paterson never traveled to America so he synthesized cartographic information from maps in circulation at the time, and a comparison of these with Paterson's reveal many exact or near similarities. One map that must have served as his basic model was Louis Delarochette's *British Dominions in North America*, published by John Bowles. Paterson stripped this map down to bare essentials. There is little inland topography and geography not essential to the cantonments of troops was simply excluded. On some examples, Long Island completely disappears as a landmass. Philadelphia, the largest American city, is excluded from the earliest example and Boston does not appear on any of them. Regiments are identified by red rectangles, while companies (typically nine to a regiment) appear as small red bars.

One characteristic of all of the cantonment maps is the presence of a Royal Proclamation Line. Just beyond it to the west was written in bold lettering LANDS RESERVED FOR THE INDIANS. Fort Pitt was located just within that line and became a critical outpost as families from Virginia began building homes beyond the line in the Monongahela River valley. Gage was somewhat sensitive to the rights of the Indians and opposed expanded colonial settlements, writing Lord Halifax on January 7, 1764, "We shall avoid Many future Quarrels with the Savages by this Salutary Measure." The Proclamation Line imposed no limitations on the Crown as Gage concluded, "I have great Reason to believe, that the Savages never set any Value upon their lands, till we taught them to Value them . . . And I am persuaded the Crown may at present obtain very Considerable Tracts of Land in both the Floridas, at a trifling Expence."

The British also sought a strong footing along the Mississippi River but were unfamiliar with the Indian tribes inhabiting that area. As early as May 21, 1764, Gage ordered Major

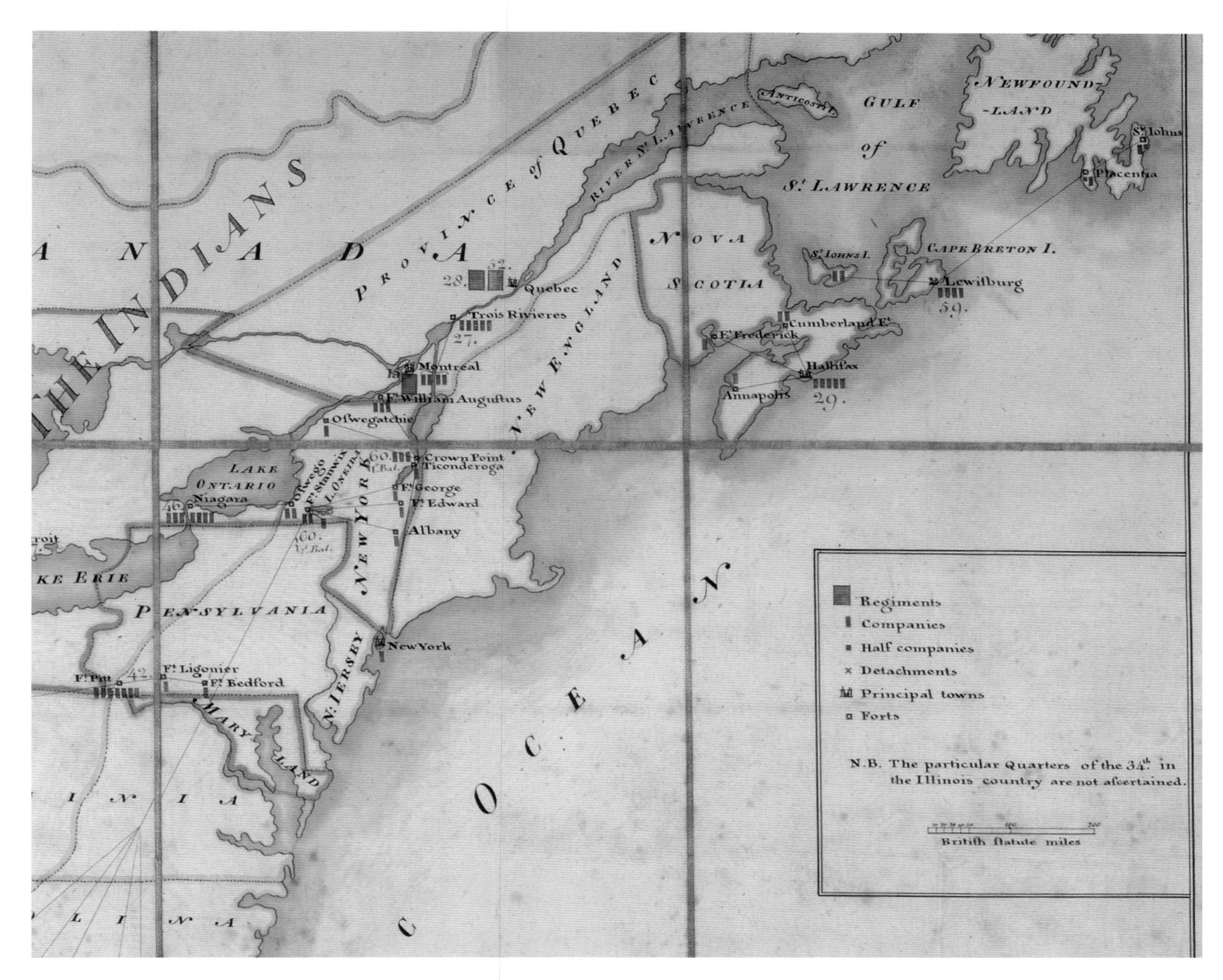

Figure 27. *Cantonment of the Forces in N. America 1766* (detail), 1766, manuscript, 51 × 62 cm. Clements Library, University of Michigan

Farmar to proceed from Mobile and "immediately endeavor to push up a Detachment of the 34th Regt to the Natchez, a Post of Consequence to prevent the French from having an Intercourse with the Numerous Nations on the East Side of the Mississipi . . . to become acquainted with Tribes to whom we are unknown. And be enabled to conciliate their Affections, and remove the Prejudices they have conceived against us." On the early cantonment maps the 34th Regiment is still scattered along the Mississippi but on later examples it has been consolidated at Fort Chartres, the last French fort ceded to the British.

The troops suffered excessive austerities at some of these cantonments. Gage reported that the 31st Regiment at Pensacola had lost almost one hundred men as a result of the heat and poor accommodations. "This Disorder has been attributed to their arriving in the extream Heat of the Season, and troubled with a bilious Fever; and being crowded in the despicable, confined Hutts of the Garrison, there being no proper lodgements erected for the Troops with the want of a regular Hospital." Gage attributed the problems of the 31st as well as the 21st Regiment in Mobile to a terrain that was "very unhealthy, being Situated in a wet Soil, near the Conflux of the fresh and Salt Waters."

The Library of Congress cantonment map shows the "General Distribution" of troops ordered by Gage dated March 29, 1766, in red, "with the alterations to summer 1767 done in yellow." These actions were well under way on June 24, 1766, when Gage wrote Secretary of State Henry Conway, "Troops are now in Motion in this Country, Six Companys of the first Battalion of the Royal American Regiment are Marching out of this Province to the City of Quebec, The Second Battalion of the said Regiment is moved towards the Lakes, to relieve the 17th Regiment in the distant Forts; and will be posted two Companys at Missilimakinak, three at Detroit, Three at Niagara, and one at Fort Ontario. The 46th Regiment is lately arrived here [New York City] from Niagara and I soon expect the 28th from Canada." The map vividly shows Paterson's efforts to stay abreast of the cantonments, which were changing rapidly from those portrayed on the early 1766 Clements map, a detail of which appears in figure 27.

Gage's letters to his secretaries are notable documents of a conscientious and capable military administrator who directed every troop movement taking place in America. They give the appearance of a commander in control of the unwieldy spoils of the greatest war that had been fought up to that time. Yet, as Gage's biographer John Richard Alden observed, "Although there was general agreement among British politicians regarding the size of the army in America after 1763, there was none regarding its disposition." The precision of Paterson's maps belies the impact of waffling politicians in London and the months required for communications between the British capital city and New York. Notwithstanding these shortcomings, the maps of the cantonment of forces in America provide a remarkable visual record that brings Gage's letters to life. They chronicle some of the early miscalculations that led to the downfall of the British empire in America.

The Stamp Act Crisis

By late 1765 General Thomas Gage's troop deployments in the American West had resulted in an uneasy peace on the sparsely inhabited frontier. Along the coast of the thirteen colonies, however, crisis loomed. The French and Indian War had brought England to the brink of bankruptcy, and the British began to impose taxes on the colonies in an effort to share the cost of an army to maintain the peace in America. With France defeated, the colonists saw little need for such an expense, and the imposition of taxes had an incendiary effect. "If Taxes are laid upon us," Samuel Adams of Massachusetts wrote in 1764, "in any shape without our having a legal Representation where they are laid, are we not reduced from the

Character of free Subjects to the miserable State of tributary Slaves." Under the banner of Sons of Liberty, resistance to taxation metamorphosed from words to action.

The Stamp Act that Parliament passed on March 22, 1765, required many printed materials and legal documents issued in the colonies to be written on costly stamped paper. Riots against the act erupted first in Boston and soon discontent spread throughout the colonies. A Stamp Act Congress was convened in New York City during October 1765—the first gathering of elected officials from the American colonies to oppose a British act. Although the congress put forth a list of rights and grievances, riots broke out in the city before any of them could be addressed.

John Montresor, the important British engineer and mapmaker, recorded the events of October 23, 1765, in his diary. "Arrived the Vessel with the Stamps, conducted by the *Coventry* and *Garland* Frigates–2000 people (mob) on the Battery expecting the Stamp would be landed, but were disappointed . . . they were secretly landed in the Fort [George] and took charge of by the Governor." A single regiment of a hundred men commanded by Artillery Major Thomas James defended Fort George, as shown on the detail of the cantonment map at the Clements Library (figure 27). During the next week, the governor and the stamps were confined in the fort by the increasingly hostile mob. There were "Many Placards put up threatening the Lives, Houses, and properties of anyone who shall either issue or receive a stamp." Major James himself was "threatened to be buried alive by the Populace as commanding the troops in the Fort for the protection of the Stamps."

On the morning of November 1, the day the act was to go into effect, five thousand angry citizens took to the streets of New York. Many threw stones at the fort and taunted the soldiers, while others roamed the city destroying property. Major James's fine home and all its contents fell victim to this mob.

On November 3 the British seemed resigned to the imminent storming and capture of the fort: "Obliged to spike our Guns on the Battery & also the Ordnance Guns in the Artillery yard." On the following day, General Gage wrote Secretary Conway about the futility of his small force defending the fort. "The people idle, and exasperated, the whole wou'd immediately fly to Arms, and a Rebellion began without any preparations against it, or any means to withstand it." Gage desperately needed a face-saving solution. "The mob it seems intended . . . to collect the Comm' in Chief," wrote Montresor, "also the friends to the Government and have marched them in front when . . . attacking the Fort." At the peak of the crisis, agreement was reached to hand the stamped papers over to the Corporation of New York for safekeeping. The city would not issue the stamped paper but promised to reimburse the Crown for its value. Although Gage had been forced to capitulate to the mob, he had deferred the revolution.

During the Stamp Act crisis, the unfolding events were being watched from three vessels of the Royal Navy patrolling the New York harbor. As the capitulation of the fort appeared imminent, the governor's family sought protection on one of these ships, the *Coventry*. In addition to Governor Cadwallader Colden was his son, Alexander, the surveyor general of New York. Alexander was very likely on board with his senior surveyor and mapmaker, William Cockburn, who produced the remarkable cartographic view of New York dated November 1, 1765, and shown in figure 28. Archibald Kennedy, commander

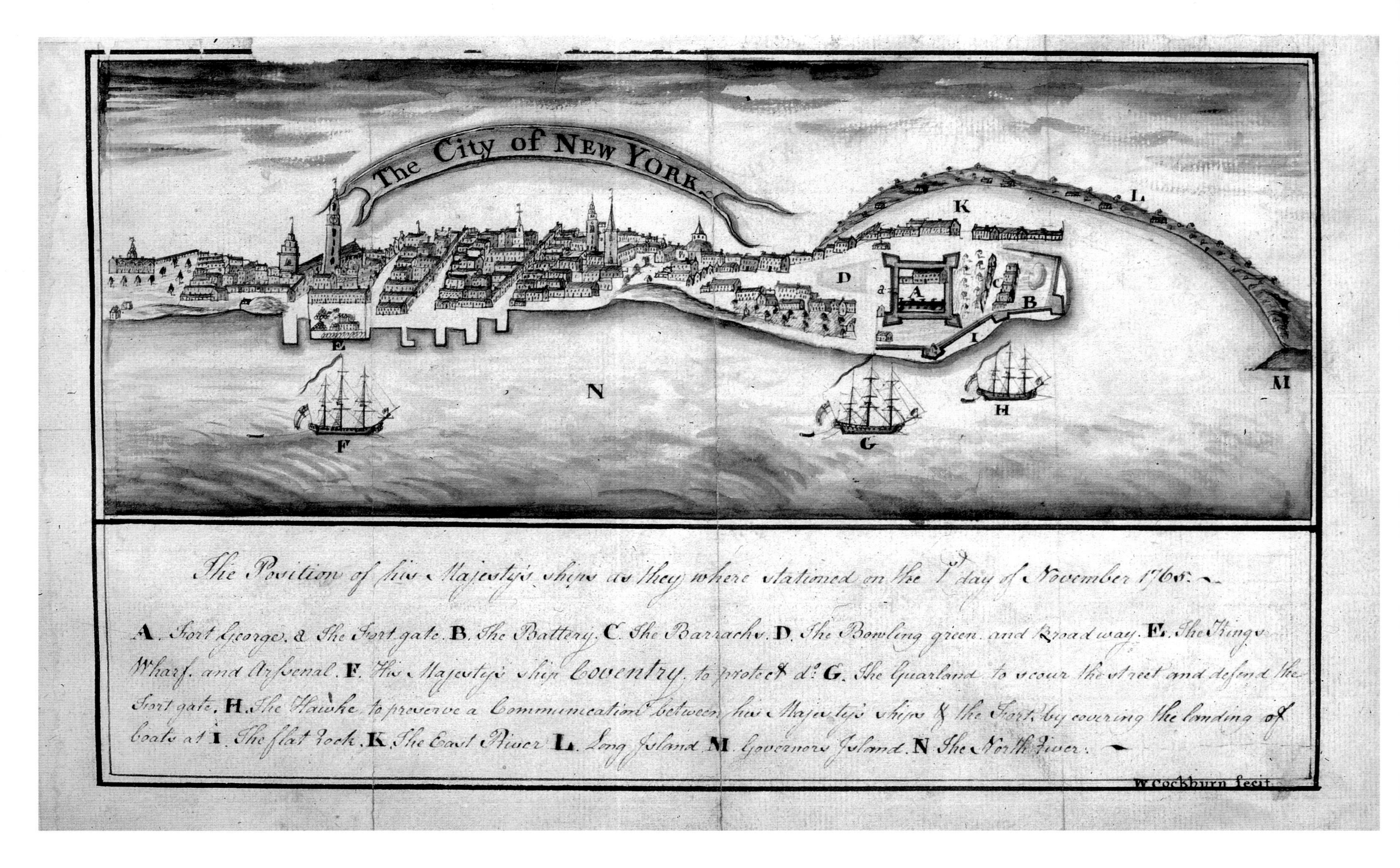

Figure 28. *New York. A Perspective View Across the North River, showing the positions of HM Ships on 1 November 1765*, William Cockburn, 1765, manuscript, 20 × 32 cm. The National Archives, Kew

of the *Coventry* (F), carried it back to London, along with a letter describing the events. Prominently displayed are Fort George (A) and The Kings Wharf and Arsenal (E). Gage feared that further inciting the crowd would lead to an attack on the arsenal where military provisions were stored.

William Cockburn would become one of New York's most prolific surveyors and mapmakers. His family archive at the New York State Library includes twelve hundred manuscript maps, including an important 1780 *Map of New York and Parts Adjacent.* In cataloging Cockburn's manuscripts, the library notes that he "sometimes rendered in near three dimensions," as he has done here. Contemporary depictions of important events between the wars are rare. The fact that a colonial drew the map rather than Montresor or James, both prolific military mapmakers, increases its distinction. The British National Archives in London holds a vast trove of documents spanning one thousand years of British history, and this little work appears on its website as one of forty-four great treasures alongside the Magna Carta and the Domesday Book.

The success of the agitators inspired them to greater heights. The events were "so much beyond riots," wrote Gage to the Viscount Barrington on January 16, 1766, "and so like the forerunners of open Rebellion, that I have wanted a pretence to draw the troops together from every post they cou'd be taken from." A little more than two months later, General Gage addressed the situation by recalling two full regiments, the 46th and the 28th, from Fort Niagara and Quebec to New York City. These actions and others appear on the map in figure 25 (page 52). The Stamp Act was repealed on March 18, 1766. While Gage did not learn of this until after he had ordered the disposition of troops on March 29, it probably would not have altered his decision. Gage had become convinced of the rebellious nature of the colonies, and the colonials had become deeply mistrustful of Parliament. There is no known copy of Daniel Paterson's map updated for 1768, but if he made one it would have shown new cantonments in Boston. Here, British troops were not as constrained as they had been in New York. The Boston Massacre in 1770 would set in motion a chain of events that would first lead to open rebellion and then to the American Declaration of Independence.

pages 60–61: Detail, figure 64. *Plan de la Ville, Port et Rade de Newport*, Louis-Alexandre Berthier, 1781, p. 127

PLAN DE LA VILLE, PORT ET RADE DE NEW_
OCCUPÉE PAR L'ARMÉE FRANCAISE AUX ORDI
L'ESCADRE FRANCAISE COMM

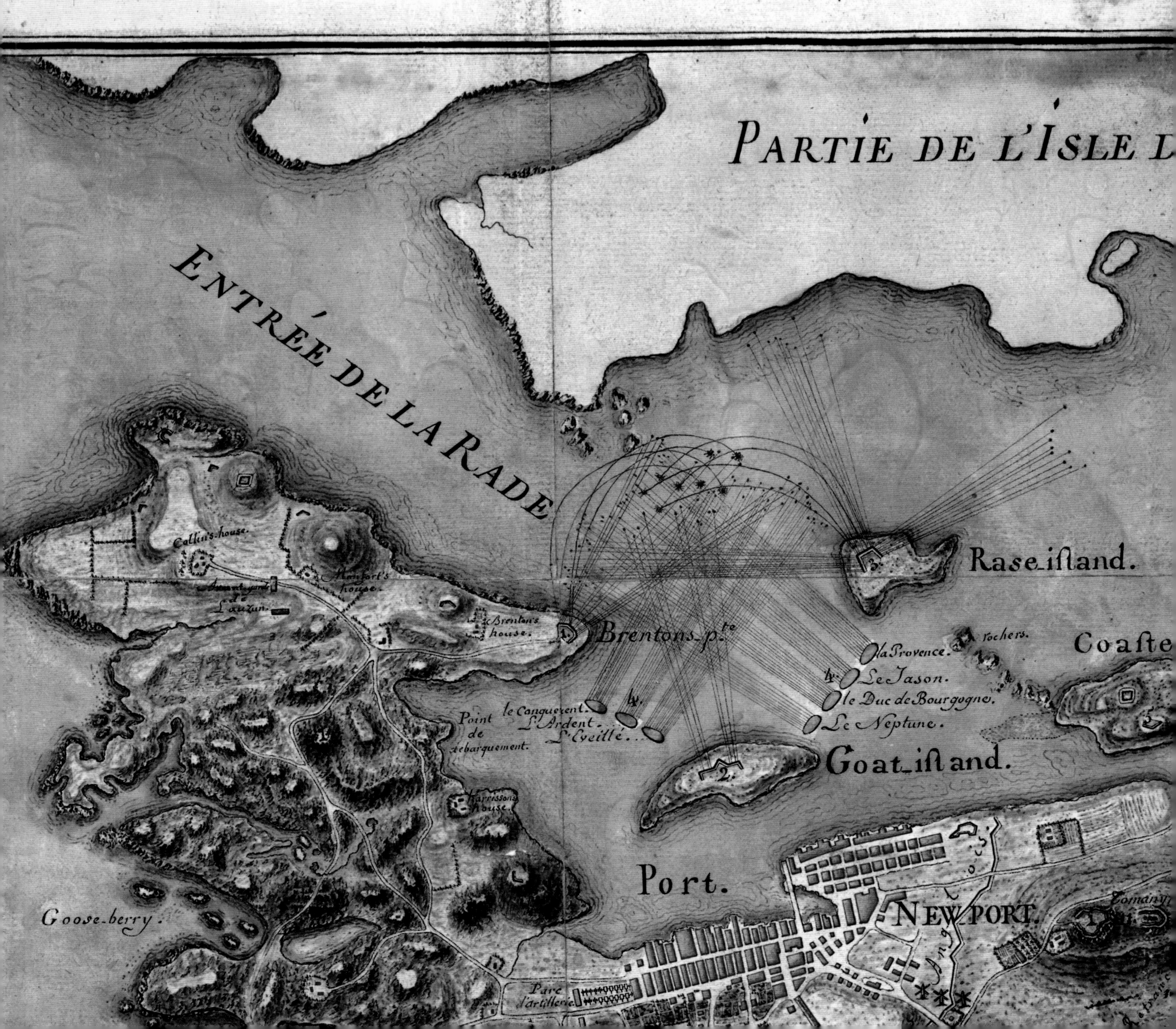

RT, AVEC UNE PARTIE DE RHODE-ISLAND
DE MR. LE COMTE DE ROCHAMBEAU ET DE
DÉE PAR MR. LE CHR. DESTOUCHES.

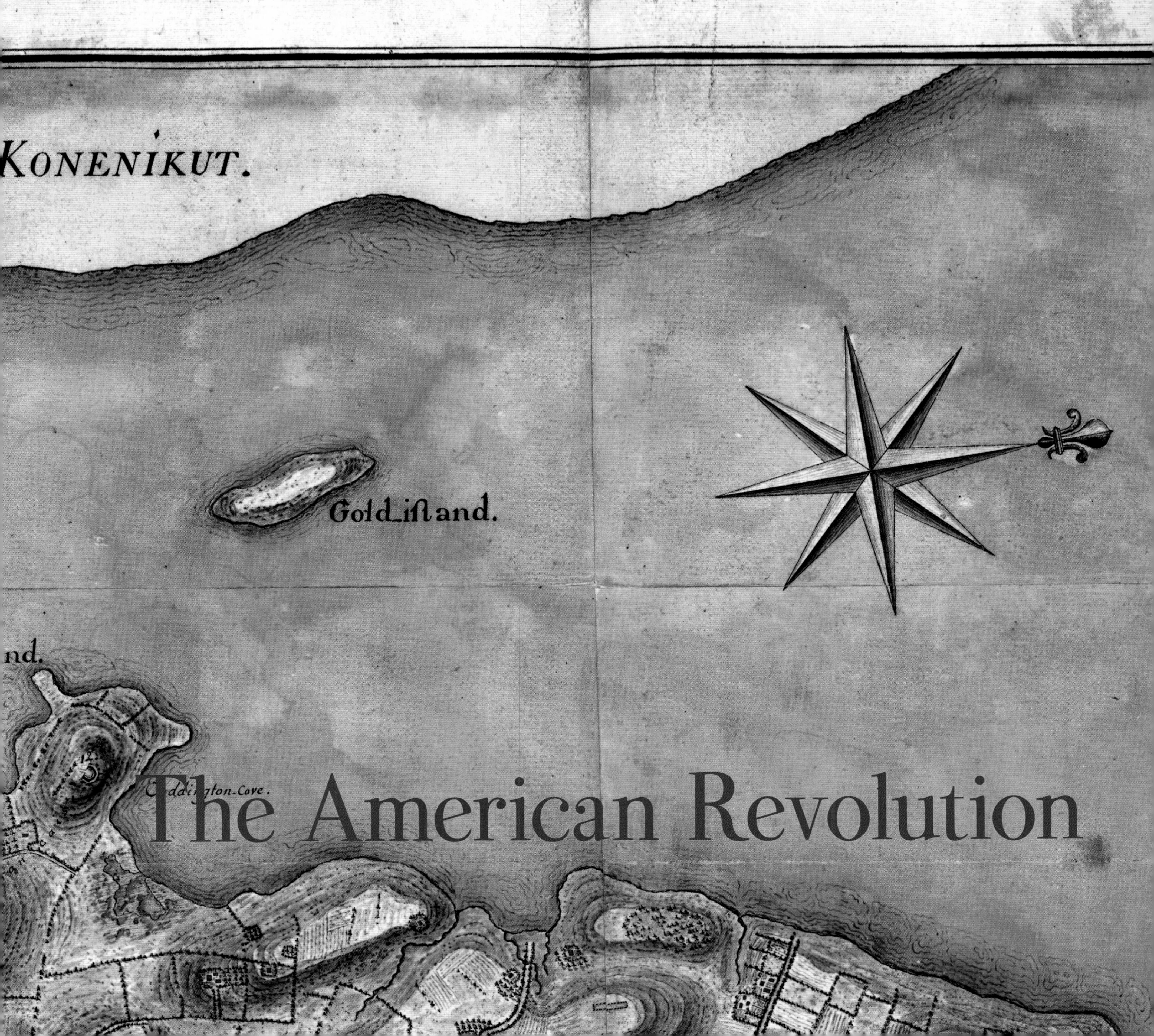

The American Revolution

Minutemen
Woobum
Minutemen
Medford
Lexington
Maldon
Lord Percy's return from Lexington
Col. Smiths return from Concord
Navigable to the Bridge for Vessels of 60 Tons
Mistick R.
Winter Hill
Magazine
Temples Farm
Penny Ferry
Monatomy
Gen.l Putnams Camp
Winnsimit
Provincials firing behind the Walls
Part of Winter Hill
Charles Town
Ferry Boat
Cambridge
Noddles I.
Head
Quarters
Phipps's Farm
Waltham
Bever Brook
PLAIN
Water Town
Charles R.
BOSTON
Bridge
WATER TOWN
Road to Worcester
Brookline
Castle W.m
Water Town Hill
Dorchester Neck
Principal part of the Provincial Army as encamped commanded by Gen.l Washington
Roxbury Hill
Dorchester
Weston
JAMAICA PLAIN
Jamaica Pond
Gen.l Ward's Camp
Thompson
Falls
Falls
M.r Walters
Meeting House
Milton R.
Dedham

Lexington and Concord

To see the earliest plan of the engagements that gave rise to the Revolutionary War in America, a trip to the village of Alnwick in northern England is in order, where the bulk of a medieval castle looms large over the moorland. After Windsor, Alnwick is the largest inhabited castle in all of England. Its gloomy antiquity made it the perfect setting for the Harry Potter films, where some of the episodes were filmed. It is the ancestral home of the dukes of Northumberland, and of the second duke, Lord Hugh Percy (figure 29), an officer and active combatant during the early engagements of the war. Percy loved maps and much of his world-class collection is on file in the archives of the castle (some have been sold at auction but not his Revolutionary War maps), including the revealing manuscript illustrated here of the skirmishes at Lexington and Concord (figure 30). The beginning of the American War of Independence is usually dated April 19, 1775, and Percy's map of the events of that day is the first to show not only the action in the two towns but also the British retreat under his command. When Percy returned to England, his luggage included a black box filled with maps delineating his battles in America. They remained undisturbed and unknown in the castle for nearly two hundred years, until his descendant the tenth Duke of Northumberland quietly released images of them to the world in 1969.

The manuscript map is not signed or otherwise identified, but a version published in London on July 29, 1775, is by one I. De Costa (figure 31). "De Costa's map is the first graphic document to depict the actions at Lexington and Concord," wrote Kenneth Nebenzahl in the *Atlas of the American Revolution*. "It is the only map issued at the time to show the marches of British forces and sites of the major skirmishes." The manuscript was no doubt the work of De Costa as well. Unfortunately, aside from this map, nothing is known of the mapmaker other than his probable collaboration on it with Jonathan Carver, the explorer of the Great Lakes and the Mississippi.

Discontent in America had reached a high pitch on that day in April when both the British and their disquiet colonists toughened themselves for a moment that had been in the works for months. "Things now every day begin to grow more and more serious," Lord Percy wrote home on April 8. He had been stationed in Boston during the winter of 1774–75, as patriot units had been raised and drilled by the loosely organized group who called

Detail, figure 31, p. 68

Figure 29. *Portrait of Hugh Percy, Second Duke of Northumberland*, Gilbert Stuart, ca. 1788, oil on canvas, 74 × 62 cm. High Museum of Art

themselves Minutemen. One of the biggest obstacles facing the colonists as they readied themselves was the basic matter of provisions. Gunpowder was in short supply (in fact, the exportation of gunpowder to Massachusetts had been banned in the fall of 1774) and firearms were rudimentary and few in number. Thomas Gage, who in 1763 succeeded Jeffrey Amherst as commander in chief of the British forces in America during the French and Indian War, was appointed commander in chief of the military in North America and royal governor of Massachusetts in 1774. In this latter capacity he received a January 27, 1775, dispatch from the Earl of Dartmouth ordering him to "arrest and imprison the principal actors & abettors in the Provincial Congress," the leaders of the incipient rebellion. Gage's precipitous response to the letter was the pivitol act that began the war for American independence.

The opening scenes of the Revolutionary War were reported by Gage in his letter to Dartmouth of April 22, 1775. "I am to acquaint your lordship that having received Intelligence of a large Quantity of Military Stores being collected at Concord, for the avowed Purpose of Supplying a Body of Troops to act in opposition to His Majesty's Government." Under the command of Lieutenant Colonel Francis Smith and Major John Pitcairn, Gage ordered troops to "destroy the Said Military Stores" in Concord and arrest John Hancock and Samuel Adams, who were attending the Provincial Congress in Concord and spending their nights at Jonas Clarke's parsonage in Lexington. Once arrested, they were to be sent to London in chains to stand trial for treason.

Dr. Joseph Warren, a leading advocate of patriot causes, learned about the mustering and summoned William Dawes and Paul Revere to warn of the impending danger. On the evening of April 18, 1775, Revere's famous ride included a stop at the parsonage where Hancock and Adams were staying. He alerted the two men of the British activity but they did not take the warning seriously. In fact, hours passed before they fled the comfort of the well-appointed house. Hancock had things other than British soldiers on his mind that evening. He had just received the gift of a freshly caught salmon and was reluctant to abandon the perishable treasure. At dawn, Hancock finally agreed to quit Lexington, and with Adams and Revere in his coach they drove to Woburn (Burlington). That salmon, however, was never far from Hancock's thoughts and his luxurious coach—befitting the richest man in New England—was ordered back to retrieve the fish, which was subsequently cooked to perfection by his hostess.

After England had implemented the various Intolerable Acts, a Second Massachusetts Provincial Congress had convened in Concord and measures were taken to arm. In this rural town, twenty-one miles from Boston, the Minutemen established their depot of military supplies. Despite the town's small size Gage, acting on his letter from Lord Dartmouth, set his sights on Concord. This hamlet occupied strategic land and was a place where Bostonians often congregated with residents of outlying rural towns; it also became the central location to store arms and supplies, including cannon. In the days before the encounter, the colonists watched as British military forces mobilized around Boston. Soldiers began coming ashore from the transports in the harbor, which can be clearly seen lined up on Lord Percy's plan, and the colonists braced themselves for whatever was to come.

Colonel Smith was the senior officer in the Boston barracks, and Gage commanded him to lead seven hundred troops to Concord. "A Quantity of Ammunition and Provision," read the order, "together as Number of Cannon and small Armes having been collected at Concord for the avowed Purpose of asserting a Rebellion against His Majesty's Government, You will march with the Corps of Grenadiers and Light Infantry put under your Command with the utmost expedition and secrecy to Concord, where you will seize and destroy all the Artillery and Ammunition, provisions, Tents & all other military stores you can find."

Silently and in the dead of night, the British column crossed the Charles River by boat, and then hiked through Cambridge, Somerville, and Menotomy (Arlington). It was dawn when Smith's soldiers reached the Lexington Common on their way to Concord. Revere had spent the night alerting the countryside that the British military was on the move. "He waked the captain of the minutemen," wrote the historian Allen French of the famous ride, "and after that alarmed almost every house until he got to Lexington." Seventy armed Minutemen under orders from John Parker, the captain of the local company, blocked the progress of these British soldiers at the Lexington bridge.

At age fifty-one, Smith was a lumbering commander whose excessive weight made him an ample target for the enemy. He also had a reputation for being a "slow thinker." Such was the leader of the first major clash with the rebels. John Pitcairn, the other officer in the field, on the other hand, was widely admired in the British barracks. He had a reputation for instilling trust among his men and for his knowledge of the local geography. He knew the road to Concord and is said to have "studied the town in disguise." Such were the dramatis personae who met on the bridge in Lexington, Massachusetts, in April 1775.

At first there was a standoff. The adversaries glared at each other on the bridge, neither side wanting to break the deadlock. Pitcairn ordered the Minutemen to lay down their arms. Parker was in the act of complying, telling his men to disperse, but just as they were beginning to disband a shot rang out. Considerable debate has ensued as to where it came from, but to this day it is still not known who fired it. Gage, however, had no doubt about where that first shot had come from: Smith "was Opposed," he wrote to Dartmouth, "by a Body of Men, within Six Miles of Concord: Some few at first began to fire upon his Advanced Companys, which brought on a fire from the Troops, that dispersed the Body opposed to them." Two or three more shots were then fired in quick succession followed by an order from a British officer to open fire, and within a few minutes eight Americans lay dead on the ground and ten others were wounded. Half of the British impairments were to a horse: Pitcairn's horse was slightly wounded by a passing bullet and one soldier sustained a leg injury.

The ruffled British forces, apparently not fully sensing the gravity of the situation, then marched to Concord, a few miles away. There the rebels had been busy evacuating the stored military supplies: powder kegs were rolled into the woods and other military supplies were ferreted away in attics. Fully three hundred to four hundred armed rebels stood waiting on the North Bridge for the advancing redcoats. Once in the town, a search for munitions commenced but few were found. The rebels were enraged by the actions that had taken place in Lexington, and an exchange of fire took place. It lasted only a few minutes but the British had fatalities: three killed and eight wounded. The Americans lost two while three others were wounded. After the skirmish the British retreated and headed back to their barracks, the rebels in pursuit. They were "Attacked from all Quarters," wrote Gage, "where any Cover was to be found, from whence it was practicable to Annoy them; and they so fatigued with their March, that it was with difficulty they could keep out their Flanking Partys, to remove the Enemy at a Distance." The rebels were not conducting themselves according to the rules of European warfare, and the map shows them firing from behind walls, a practice viewed by the British as cowardly and lacking honor. Smith, who was wounded in this action, was clearly in need of help. Lord Percy entered the fray to reinforce Smith.

Percy had remained at headquarters while Smith and Pitcairn were engaging the rebels, but he and his fourteen hundred troops and two cannon (the cannon are carefully drawn on his plan at the top center of the map) were at the ready if reinforcements were required. On the evening of April 18, Gage ordered Percy's brigade to move out at 4 a.m. After a series of misadventures that caused several hours of delay, Percy was on the move at 9 a.m., but it was not until 2:30 that afternoon that he finally met up with Smith's bedraggled troops, who were still under siege. Percy managed to rally Smith's troops and escorted them back to headquarters. "Notwithstanding a continual Skirmish for the Space of Fifteen Miles," wrote Gage, "receiving fire from every Hill, Fence, House, Barn &c His Lordship kept the Enemy off and brought the Troops to Charles Town, from whence they were ferryed over to Boston. Too much praise, cannot be given to Lord Percy." Sixty-five British soldiers had died that day, including fifteen officers, and 173 men were wounded; another twenty-six were unaccounted for. Casualties on the American side were somewhat less: forty-nine dead and forty-six missing and wounded.

The map reveals the salient localities as the British troops returned from their mission to the two rural towns. At the upper left corner is identified the "Bridge where the attack began." The mapmaker committed a few errors as he traced the steps of the retreating army under Percy. As the historian William Cumming pointed out, Percy "relieved Smith at Lexington, not west of it, as on this map; he took the

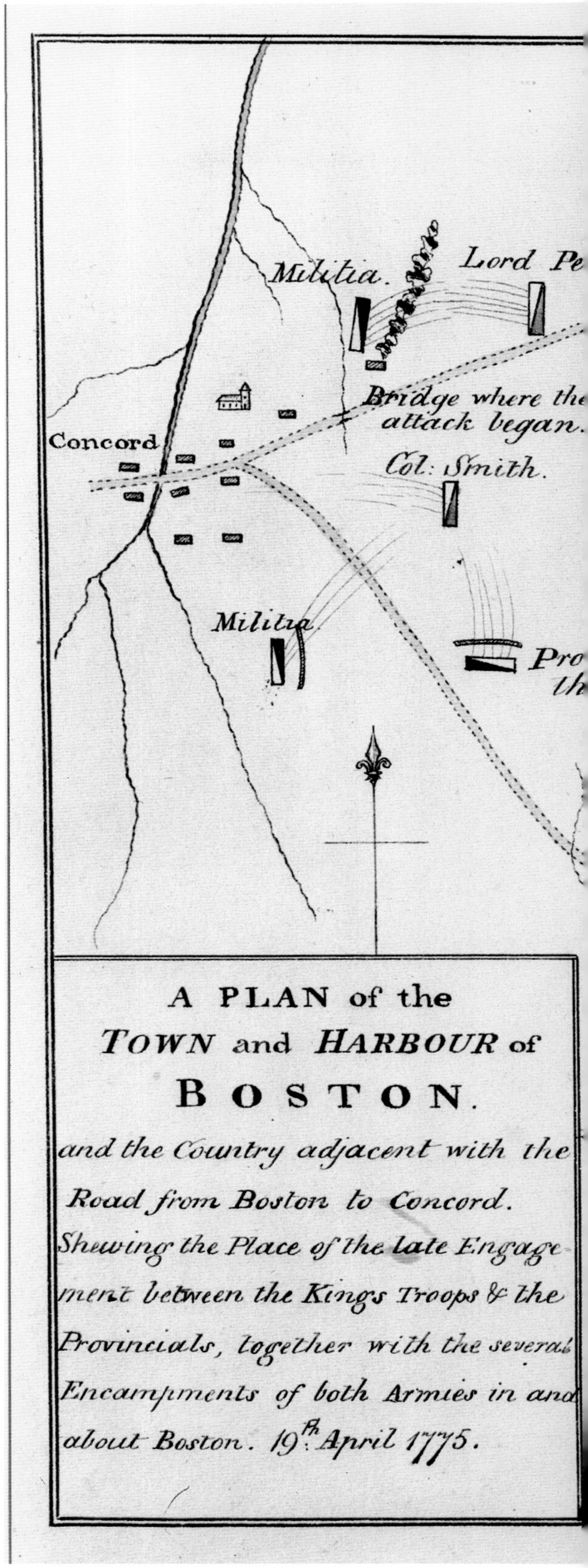

Figure 30. *A Plan of the Town and Harbour of Boston . . . Shewing the Place of the late Engagement between the King's Troops & the Provincials*, cartographer unknown, 1775, manuscript, 20 × 30 cm. Collection of the Duke of Northumberland, Alnwick Castle

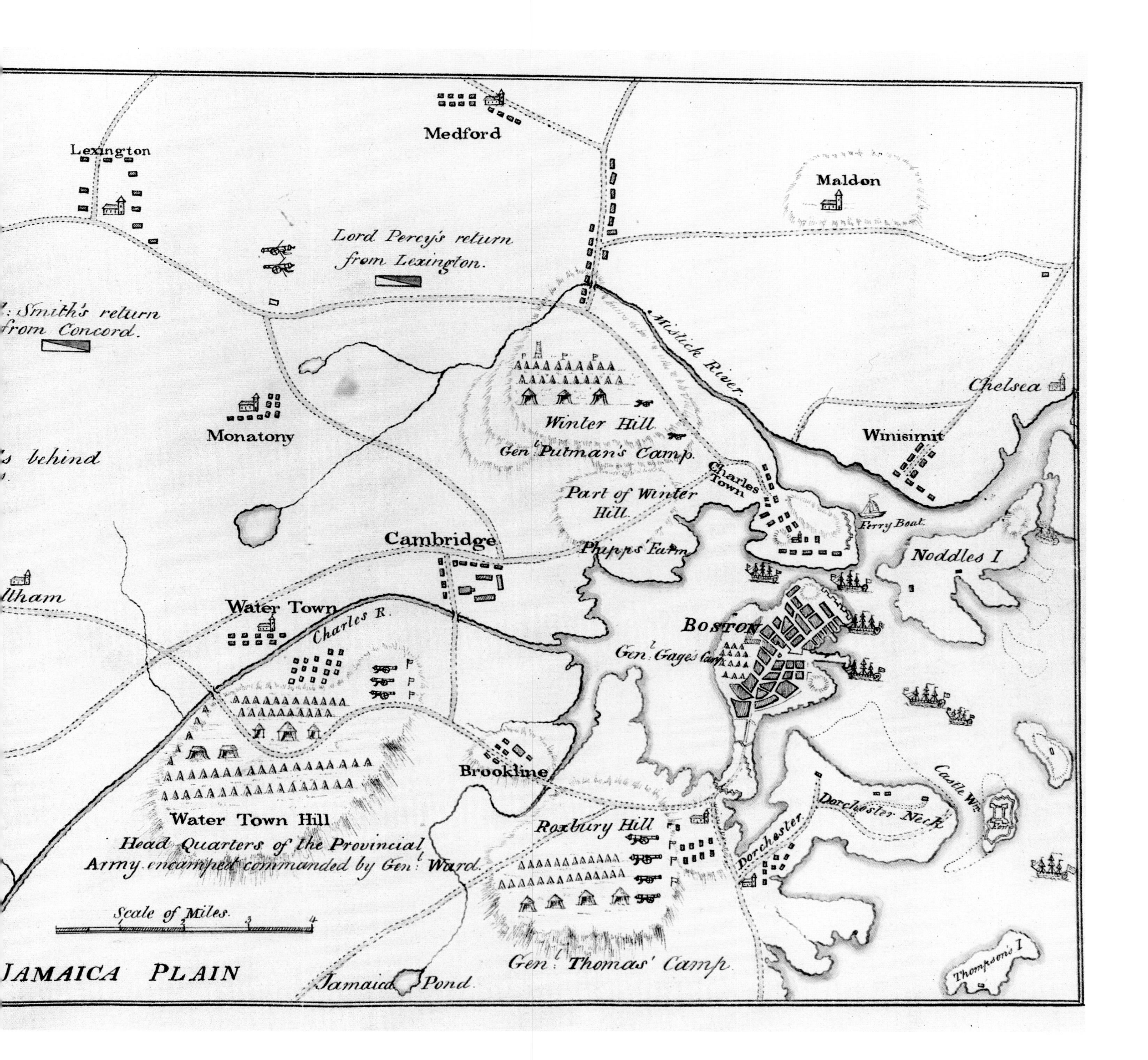

Lexington
Medford
Maldon
Lord Percy's return from Lexington.
Smith's return from Concord.
Mistick River
Monatony
Winter Hill
Gen.l Putnam's Camp.
Chelsea
Winisimit
behind
Charles Town
Part of Winter Hill.
Ferry Boat.
Cambridge
Phipps' Farm
Noddles I
Water Town
Charles R.
BOSTON
Gen.l Gage's Camp
Brookline
Castle Wm.
Dorchester Neck
Water Town Hill
Head Quarters of the Provincial Army encamped commanded by Gen.l Ward.
Roxbury Hill
Dorchester
Scale of Miles.
1
2
3
4
JAMAICA PLAIN
Jamaica Pond
Gen.l Thomas' Camp
Thompson's I

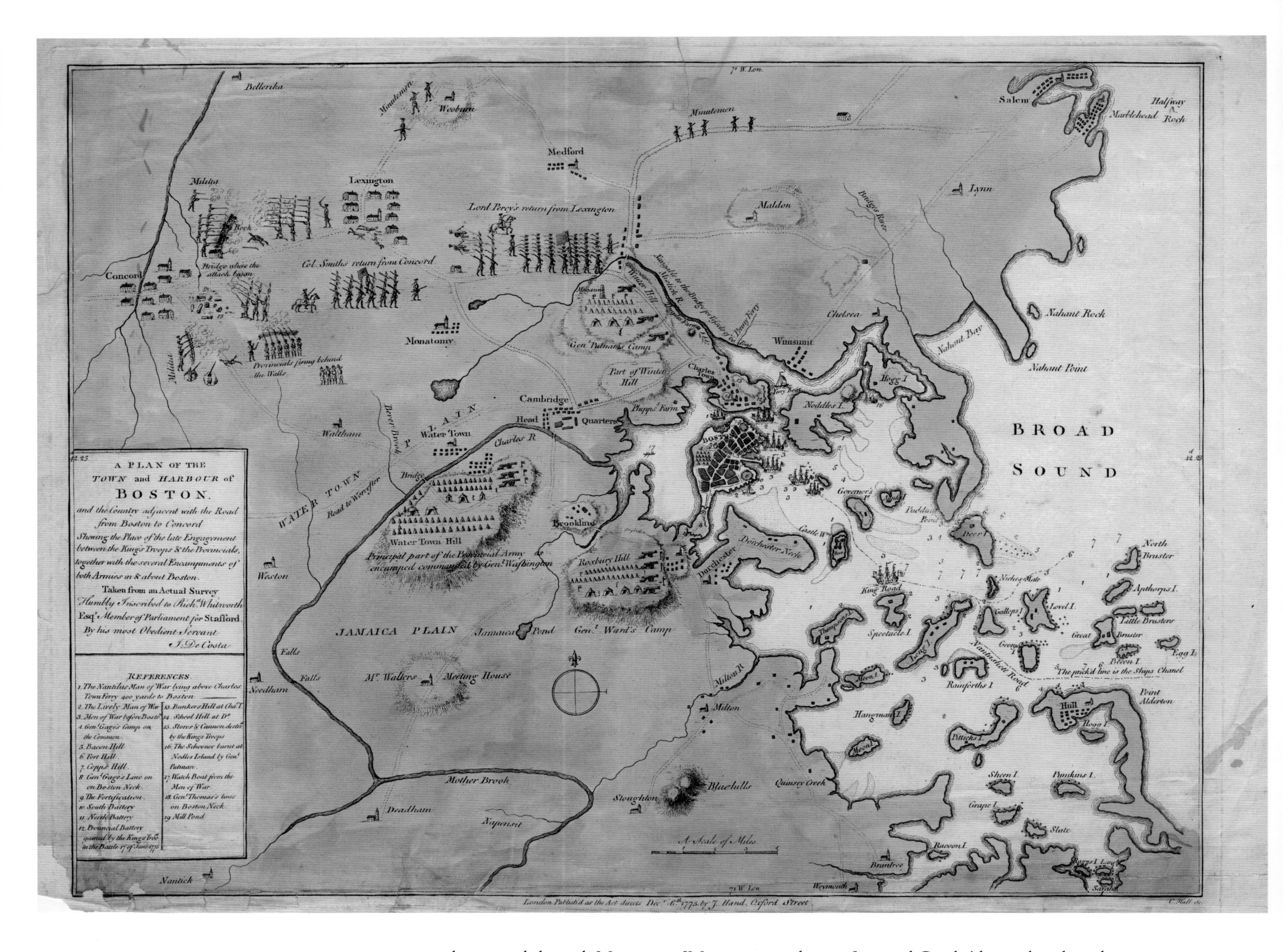

Figure 31. *A Plan of the Town and Harbour of Boston . . . Shewing the Place of the late Engagement between the King's Troops & the Provincials*, I. De Costa, London, 1775, copperplate engraving, 37 × 49 cm. The Newberry Library

lower road through Menotomy [Monatony on the map] toward Cambridge, rather than the upper fork, as the plan's author wrongly shows." Several British campgrounds are shown on the map on the fringes of the troubled landscape.

When the printed version of the map appeared in London more than three months after the skirmishes, the Battle of Bunker Hill had already been fought and the gravity of the situation was clear to the British. The printed map (above) even included references to Bunker Hill. Lord Percy's manuscript is not signed, but the printed version is by I. De Costa. Much time and labor has been expended by map historians to learn something about the maker of the first map of the American Revolution, but nothing so far is known. Not even his first name. Yet it remains the only map contemporaneous with the events to show the action on the day His Majesty's army marched against his rebellious colonists.

Bunker Hill

John Montresor (figure 32) was the most able British soldier who served during the French and Indian War and the American Revolution. His tour of duty in America lasted twenty-three years, from 1755 to 1778–with one exception, longer than any other British engineer. "He had greater ability than any of the generals under whom he served," wrote the historian Kenneth Roberts of Montresor, "and if he had commanded British troops in America, the Revolution would have ended in 1776." Montresor was at Lexington and Bunker Hill and on Long Island as well as in Philadelphia, and some of the finest maps of the Revolution were drafted by him. He drew as many as two thousand maps while in America and many were published. Illustrated here (figure 33) is his *Draught of the Towns of Boston and Charles Town and the Circumjacent Country Shewing the Works Thrown up by his Majesty's Troops and also Those by the Rebels during the Campaign 1775.* This manuscript map from the time of the Battle of Bunker Hill was made for Lord Percy and is still in Alnwick, his vast castle in Northumberland.

Figure 32. *Colonel John Montresor,* John Singleton Copley, ca. 1771, oil on canvas, 76 × 63 cm. Detroit Institute of Arts, USA Founders Society Purchase, Gibbs-Williams Fund/Bridgeman Images

After the encounters at Lexington and Concord (April 19, 1775) the British and the rebel colonists were on an accelerating collision path. As tensions in Boston increased exponentially, there was the call to the surrounding colonies for troop reinforcements. The military population in the city began to swell as companies arrived in Boston from New Hampshire and Rhode Island as well as a regiment from Connecticut under the command of Brigadier General Israel Putnam. One factor in the looming confrontation was obvious to both sides: whoever controlled the high ground on either nearby Dorchester or the peninsula of Charlestown would hold the strategic advantage. On June 13 intelligence reached the rebels that the British intended to occupy Dorchester Heights, and on June 15 the Massachusetts Committee of Safety decided to make a countermove by occupying Bunker Hill, an elevation on Charlestown, the peninsula directly across the water from Boston.

General John Burgoyne described the military circumstances in a letter to his nephew Lord Stanley: "Boston is a peninsula, joined to the main land only by a narrow neck, which on the first troubles Gen. Gage fortified; arms of the sea, and the harbour, surround the rest: on the other side one of these arms, to the North, is Charles-Town (or rather was, for it is now rubbish), and over it a large hill, which is also, like Boston, a peninsula." Although the ensuing battle is named after Bunker Hill, most of the fighting took place on Breed's Hill. This elevation was nearer to Boston but not as high as Bunker Hill and was strongly preferred by Putnam; it would not take the rebels long to discover that this eminence did not possess as strong a tactical position.

On the evening of June 16, 1775, a patriot force of twelve hundred under the leadership of Colonel William Prescott had silently marched from Cambridge and crossed that narrow neck of land onto the peninsula of Charlestown where they built a square redoubt on Breed's Hill (each side measuring about 45 yards) and then braced themselves for the British to respond. In the early morning light of the next day, an adumbration of the redoubt came

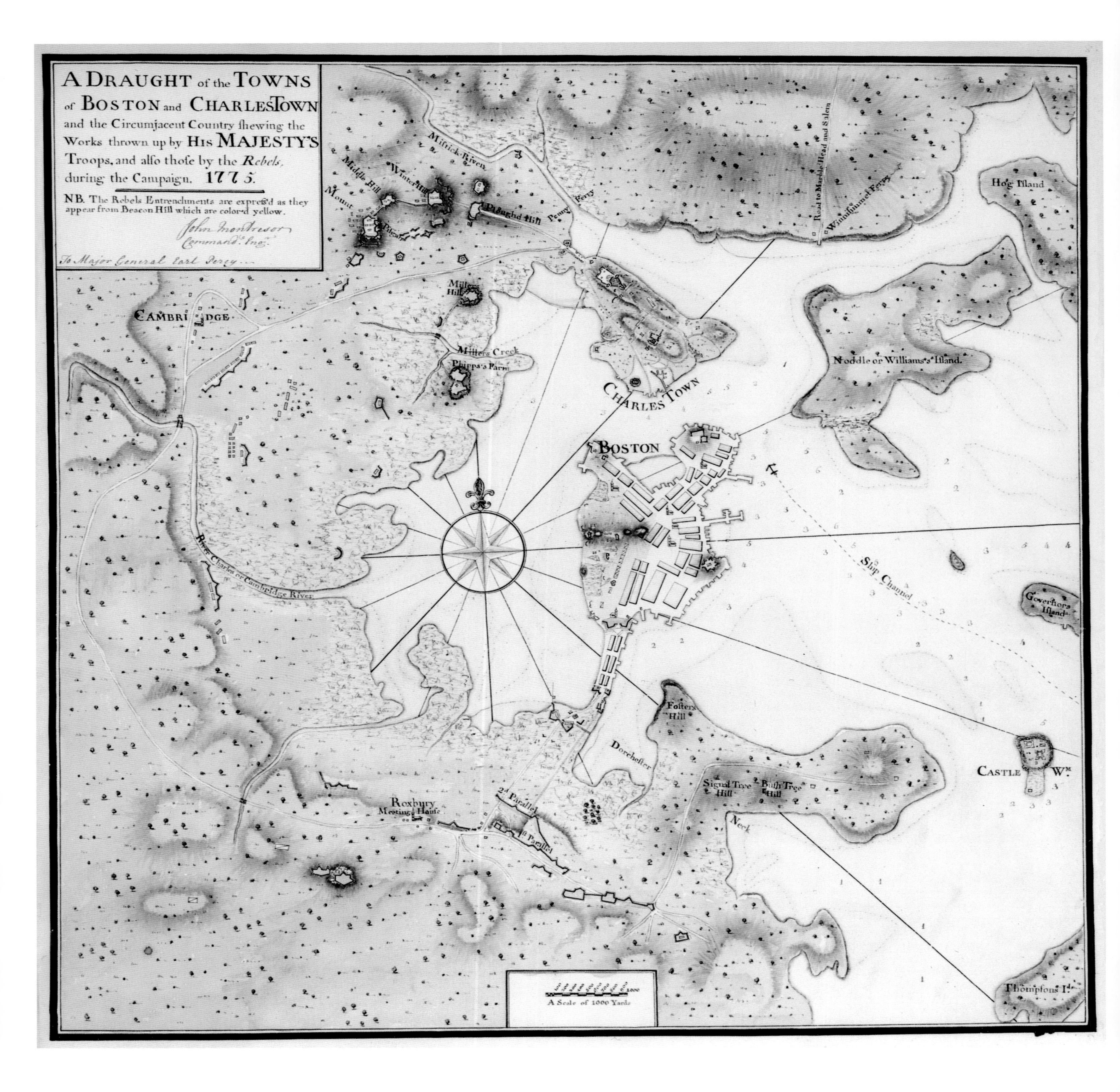
A DRAUGHT of the TOWNS of BOSTON and CHARLESTOWN and the Circumjacent Country ſhewing the Works thrown up by HIS MAJESTY'S Troops, and alſo thoſe by the Rebels, during the Campaign. 1775.
NB. The Rebels Entrenchments are expreſs'd as they appear from Beacon Hill which are color'd yellow.
John Montresor
Commandt. Engr.
To Major General Earl Percy...
Miſtick River
Winter Hill
Middle Hill
Mount
Ploughd Hill
Penny Ferry
Road to Marble Head and Salem
Winnisimmet Ferry
Hog Iſland
Millers Hill
CAMBRIDGE
Millers Creek
Phipps's Farm
Noddle or Williams's Iſland.
CHARLES TOWN
BOSTON
River Charles or Cambridge River
Ship Channel
Governors Iſland
Fosters Hill
Dorcheſter
Signal Tree Hill
Buſh Tree Hill
Neck
CASTLE Wm.
Roxbury Meeting Houſe
2d. Parallel
1 Parallel
Thompſons I.
A Scale of 1000 Yards

into the view of the early risers on board the British sloop *Lively*, at the ready in Boston Harbor. The British had planned for this action. They had built fortifications and armaments within the city of Boston, including the Cobb's Hill battery with its eight 24-pound guns and a redoubt on Beacon Hill that was armed with two 12-pounders. John Montresor was in a good position to delineate these works on his maps for he had designed many of them himself. They are accurately located on his map, as is the "Circumjacent Country" around Boston Harbor. While Montresor's map shows the general battleground, it is a manuscript map in the Faden Collection at the Library of Congress that focuses on the peninsula of Charlestown and shows "the Action which happened 17th June 1775" (figure 34).

This spirited manuscript is identified on its verso as the work of Thomas Hyde Page, a British lieutenant who fought in the battle and was severely wounded. He received a grant of ten shillings a day for life as a result of the wound. Page's map shows the American and British lines and indicates the various movements made by the troops. Also noted are several British naval ships including armed transports, floating batteries, and His Majesty's ships *Glascow, Somerset*, and *Lively*, whose watchful sailors had been the first to notice the rebels' redoubt. Cannonfire from the *Lively* signaled the beginning of warfare with the American forces, but this initial attack on the peninsula did not damage the sturdy redoubt that the Americans had constructed in the darkness of a single night.

The high tide that General William Howe needed to land his two thousand British infantrymen on the peninsula did not come in until the early afternoon so his debarkation was delayed, thus providing the defenders of the hill more time to anticipate the battle. They had built, in addition to the redoubt, a defensive line of fence rails. Once on land, the British soldiers followed Howe's order to divide into two distinct units. Brigadier General Robert Pigot attacked the redoubt while the other unit, under Howe himself, attacked the area where the rail fence had been constructed. The letter A on the map shows the position of the British as they launched their first attack. It was a predictable frontal attack that did not catch the rebels off guard.

Among the problems the British faced was a simple matter of ammunition. They had transported 6-pounders to Charlestown, but the extra ammunition they had mistakenly brought was for 12-pounders and therefore useless. The infantrymen were also limited by their hundred-pound knapsacks, which slowed their progress. Many of the pictures of the battle show the peninsula in flames as the encounters took place. Indeed, Howe had ordered Vice Admiral Samuel Graves to fire "hot shot" from the ships to set the dry fields ablaze. Smoke billowed over the landscape as the two forces battled. The rebels, primarily New Hampshire men under the command of John Stark, repelled the aggressors. The retreating British troops were in a state of stunned disbelief. How could this ill-trained, undisciplined army of rebels repel the regulars of the greatest army on earth? After that first assault the rebels started to run low on ammunition, a disadvantage they frequently encountered throughout much of the war.

Shortly after retreating from Breed's Hill, the British attacked a second time. Using their remaining gunpowder as economically as possible, the rebels held their fire until the British were almost on top of them. When they were within firing range, the Americans ferociously defended their position, causing numerous casualties among the attackers.

Figure 33. *A Draught of the Towns of Boston and Charles Town and the Circumjacent Country, shewing the Works . . . by His Majesty's Troops; and also those by the Rebels*, John Montresor, 1775, manuscript, 43 × 45 cm. Collection of the Duke of Northumberland, Alnwick Castle

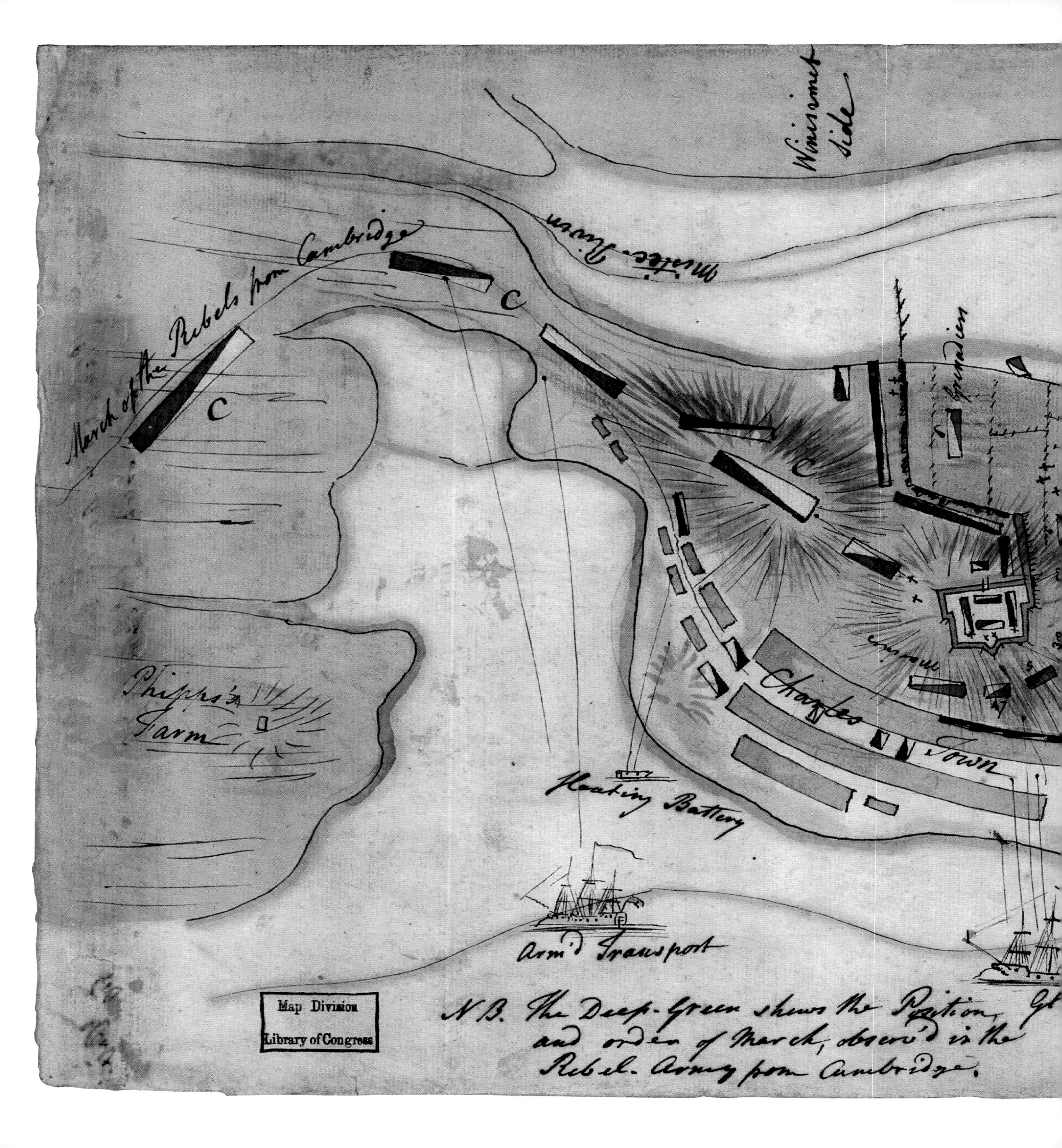
Winisimet
Side
Mistic River
March of the Rebels from Cambridge
C
C
C
Grenadiers
Charles Town
Phipps's Farm
Floating Battery
Arm'd Transport
N.B. The Deep-Green shews the Position,
and order of March, observ'd in the
Rebel-Army from Cambridge.

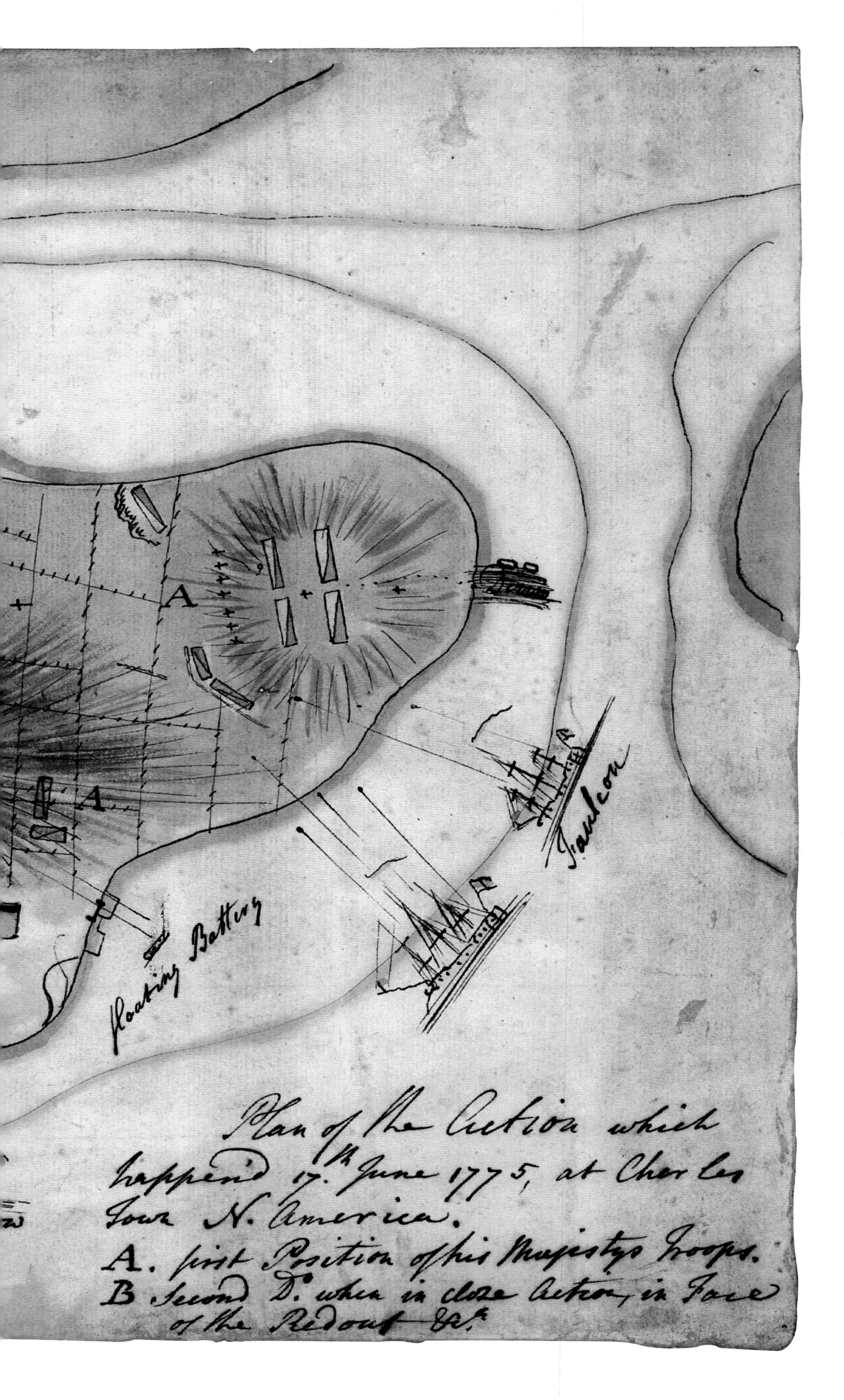

Figure 34. *Plan of the Action which Happen'd 17th June 1775, at Charles Town, N. America*, Sir Thomas Hyde Page, 1775, manuscript, 25 × 39 cm. Geography and Map Division, Library of Congress

Figure 35. *Joseph Warren*, John Singleton Copley, ca. 1765, oil on canvas, 127 × 101 cm. Museum of Fine Arts, Boston

Bayonet warfare concluded this second assault, making this retreat even bloodier than the first. "There was a moment," Howe later wrote about the battle, "I never felt before."

The British held a strong advantage when they began their third assault on Charlestown. First of all, they had four hundred fresh troops and the colonists were all but defenseless. And while the colonists had nearly run out of ammunition, the British had brought ammunition that was compatible with their field pieces. As the redcoats advanced on Prescott's redoubt, the Americans used the last of their ammunition to slow their progress. Major Pitcairn was called into service to conclude the battle, which lasted some two hours, and he was mortally wounded in this exchange. This time the Americans were forced to retreat. After fire broke out Prescott fell back from the redoubt he had built the night before. The British had held their ground and won the battle, but it was a Pyrrhic victory. They suffered over a thousand killed and wounded or nearly half of the force engaged. American casualties were estimated at 450 killed and wounded, including Dr. Joseph Warren, the president of the Massachusetts Provincial Congress (figure 35). Afterward, the British general Henry Clinton remarked in his diary that "A few more such victories would have surely put an end to British dominion in America."

Halifax

Evacuation Day is celebrated in Boston every March 17. On this day in 1776 as many as eleven hundred loyalists boarded thirty vessels in Boston Harbor and sailed to Halifax, Nova Scotia, along with fifty transports filled with nine thousand British soldiers. The British had declared themselves the winner at the Battle of Bunker Hill, but they were stunned and shaken by their devastating victory. Rather than attack again, General William Howe decided to evacuate the city.

On Evacuation Day in 1876, Bostonians celebrated the centennial of a military loss that proved to be a turning point in their favor. To commemorate the occasion, a medal that had been engraved in Paris for General George Washington was presented to the city by a civic-minded group of Bostonians (figure 36). On one side of the medal is a profile of the general himself, and on the other is a scene of Washington on horseback watching from Dorchester Heights as the British sail out of the harbor. In addition to its many historic virtues, this medal was the prototype for the Congressional Medal of Honor. It had been presented to Washington by Thomas Jefferson in 1790, and was inherited by George Steptoe Washington, the general's nephew, when Washington died in 1799. The medal was purchased from his family for $5,000 on behalf of the city and is now one of the most valuable items at the Boston Public Library.

Figure 36. Gold medal voted by Congress to General Washington (verso), courtesy of the Trustees of the Boston Public Library, Rare Books Department

When the fleet set out from Boston Harbor, Washington was not sure where the armada was headed. New York was a possibility, but Halifax was in fact the destination. This city of some five thousand in Nova Scotia became the camp for British forces in

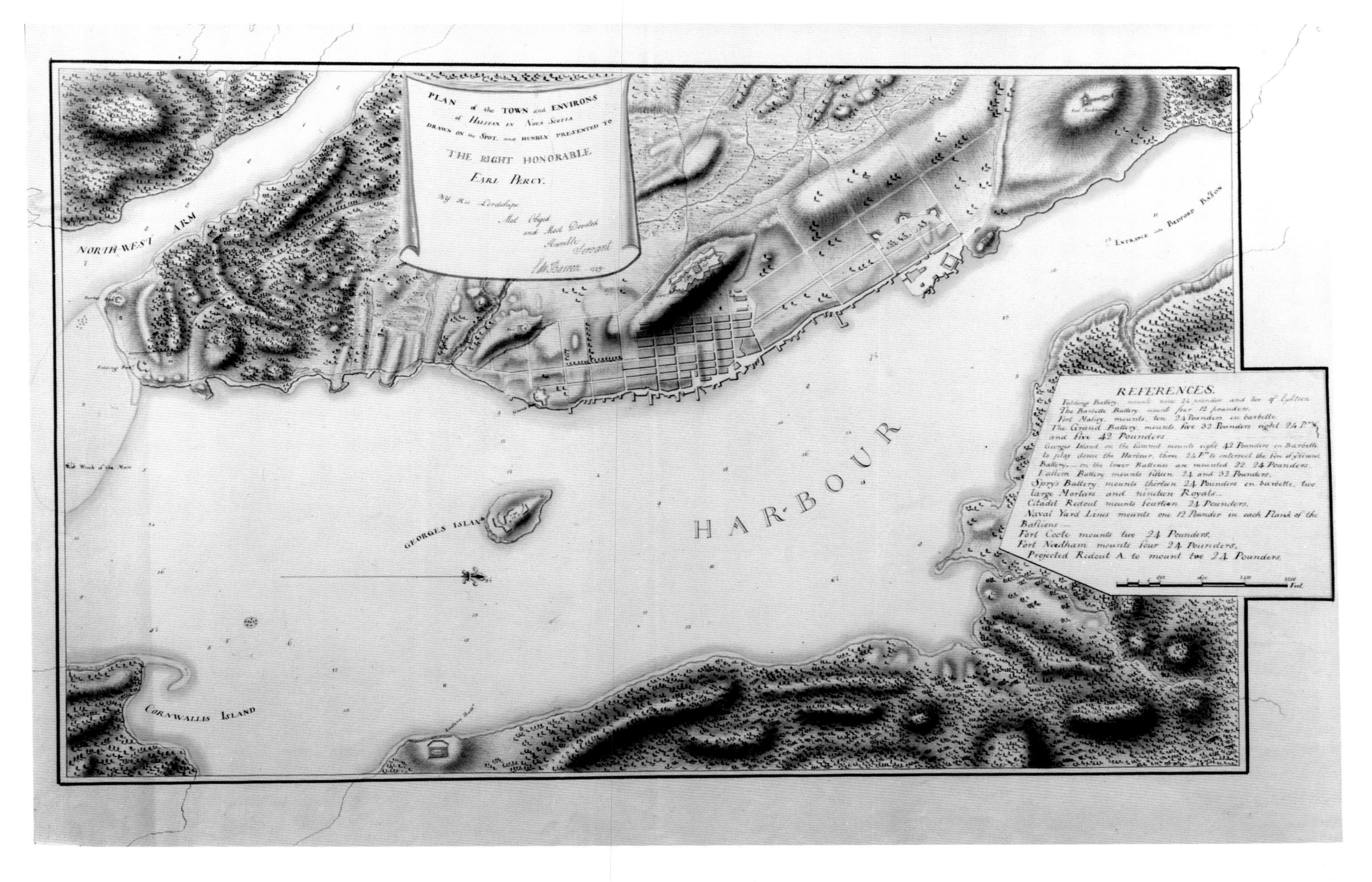

Figure 37. *Plan of the Town & Environs of Halifax in Nova Scotia*, Edward Barron, 1779, manuscript, 33 × 45 cm. Collection of the Duke of Northumberland, Alnwick Castle

America. It was an obvious place for the British to lick their wounds, await supplies and reinforcements, and plan the next course of action. It was a naval stronghold, the site of a massive fortress called the Citadel.

Plan of the Town and Environs of Halifax (figure 37) is a manuscript map presented to Lord Percy by Edward Barron in 1779, about three years after the evacuation. It shows the heavily armed harbor in a rural setting. The depths in the Halifax harbor are indicated, as is Cornwallis (now McNabs) Island. The block of references in the lower right of the map locate the installation of cannon at various sites. The Citadel is on top of the hill behind the city grid, and on top of Citadel Hill is Fort George with the imposing blockhouse at its center.

Barron, the mapmaker, was stationed at Boston in 1775 where he executed several manuscript maps for Lord Percy, including one of British troops on the Charlestown peninsula. It was drawn while the captain of the Fourth (King's Own) Regiment was convalescing from wounds received in the assault on Bunker Hill. Barron also created the Halifax oval map (figure 38) for Percy, but the date of execution is in question, as Barron does not appear to have been with the army in Halifax. Barron's regiment, however, is prominently

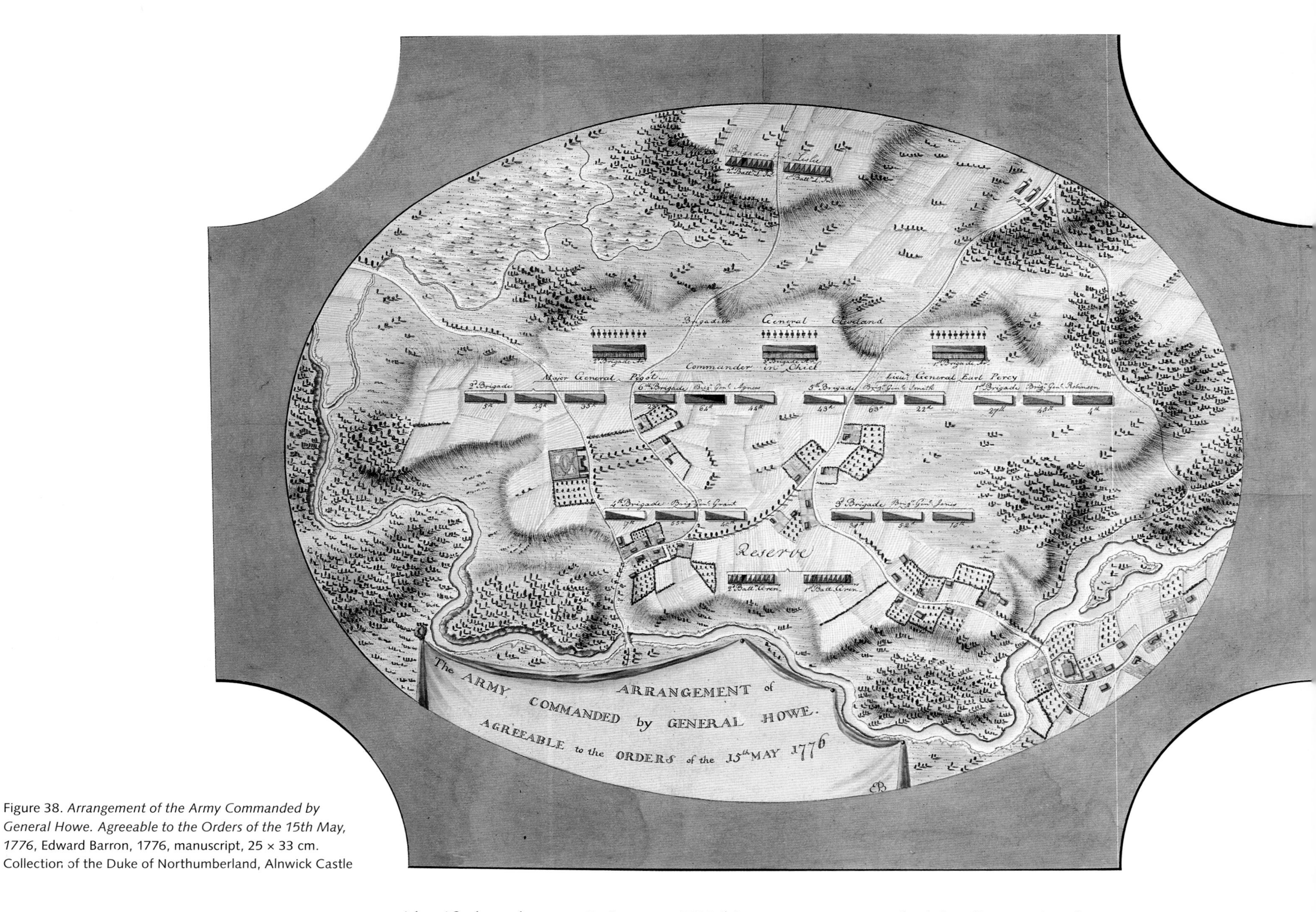

Figure 38. *Arrangement of the Army Commanded by General Howe. Agreeable to the Orders of the 15th May, 1776*, Edward Barron, 1776, manuscript, 25 × 33 cm. Collection of the Duke of Northumberland, Alnwick Castle

identified on the map. In January 1776, his company was attached (by direct orders from London) to Sir Henry Clinton and sailed to the Carolinas, seeing action at Cape Fear and Charleston before rejoining Howe at Staten Island in July.

The British overran the naval base in Halifax. "The troops filled the barracks, encamped on the Citadel slopes, and covered Camp Hill with their tents." The town was in a state of crisis during those months. "The merchants, the landlords, the brewers, the madams of the bawdy-houses reaped a harvest; but ordinary townsfolk . . . found themselves in open competition with a horde of strangers for a roof above their heads and the very food upon their tables." British troops were not idle, however, as can be seen on Barron's manuscript map, also at the Alnwick castle. It shows the *Arrangement of the Army Commanded by General Howe.* Howe tried to keep his troops battle ready by exercising them. Barron's map shows Howe's troops on the peninsula just southwest of the town as they trained on May 15, 1776, in preparation for the army's departure to New York the following month. Its precision and clarity stand in stark contrast to the chaotic, near riotous conditions the British encampment brought to Halifax as its population soared to over fifteen thousand.

The British fleet sailed from Halifax on June 9, 1776, heading for New York, and on June 29 the forty-five British ships were at anchor in the New York harbor. Within a week 130 ships under the command of General Howe's brother, Admiral Richard Howe, had arrived in the area. The ubiquitous Lord Percy was also on the scene. He had traveled with General Howe from Boston to Halifax and the two sailed from Halifax to New York, where Percy was given command of one of the three divisions of the army. It is due to Percy that the important manuscript maps illustrated here were preserved. They serve as a record of the important sojourn in Halifax. Retreat in Canada had given General Howe ample time to plan and reinforce his troops. The rebels in New York were no match for the battleships that came thundering into New York to put an end to revolution.

Charleston 1776

A month before news of Lexington and Concord had reached South Carolina, a rebellion erupted in the city of Charleston. On April 21, 1775, patriots seized stores of arms and munitions and terrorized loyalists. Britain reacted by appointing Lord William Campbell as the new (and last) colonial governor. The former navy captain vowed to quell the rebellious elements in town but soon found his life in jeopardy and his property at risk. He fled in terror to a British warship. The following year, Campbell would play an important role in a British attempt to reclaim America's fourth largest city. It would be "the first serious contest in which ships took part in this war," said the historian A. T. Mahan, and the action was captured in a detailed manuscript map and rare set of printed views, exhibited here.

In Boston, during the late autumn of 1775, General William Howe's army and navy retired into winter quarters, ignoring the Continental Army's menacing presence in nearby Cambridge. As Howe plotted the following year's campaigns against New York and Newport, he wondered if some immediate military progress might be possible in the South. In particular, he eyed the port of Charleston, a significant entry point for rebel military supplies. Howe believed that the many loyalists who lived there would support any action taken against it. General Henry Clinton was put in charge of the campaign.

News of Clinton's plans soon appeared in print. "It appears, the British Ministry and their agents . . . are preparing to make attacks upon Charleston," reported the *Proceedings of the Continental Congress* on January 1, 1776, and preparations should be made for a "vigorous defense." Washington ordered his top general, Charles Lee, to counter the British invasion. Lee knew Clinton well, having served under him in Europe. "It was a droll way of proceeding," Lee wrote about his former commander, "to communicate his full plan to the enemy."

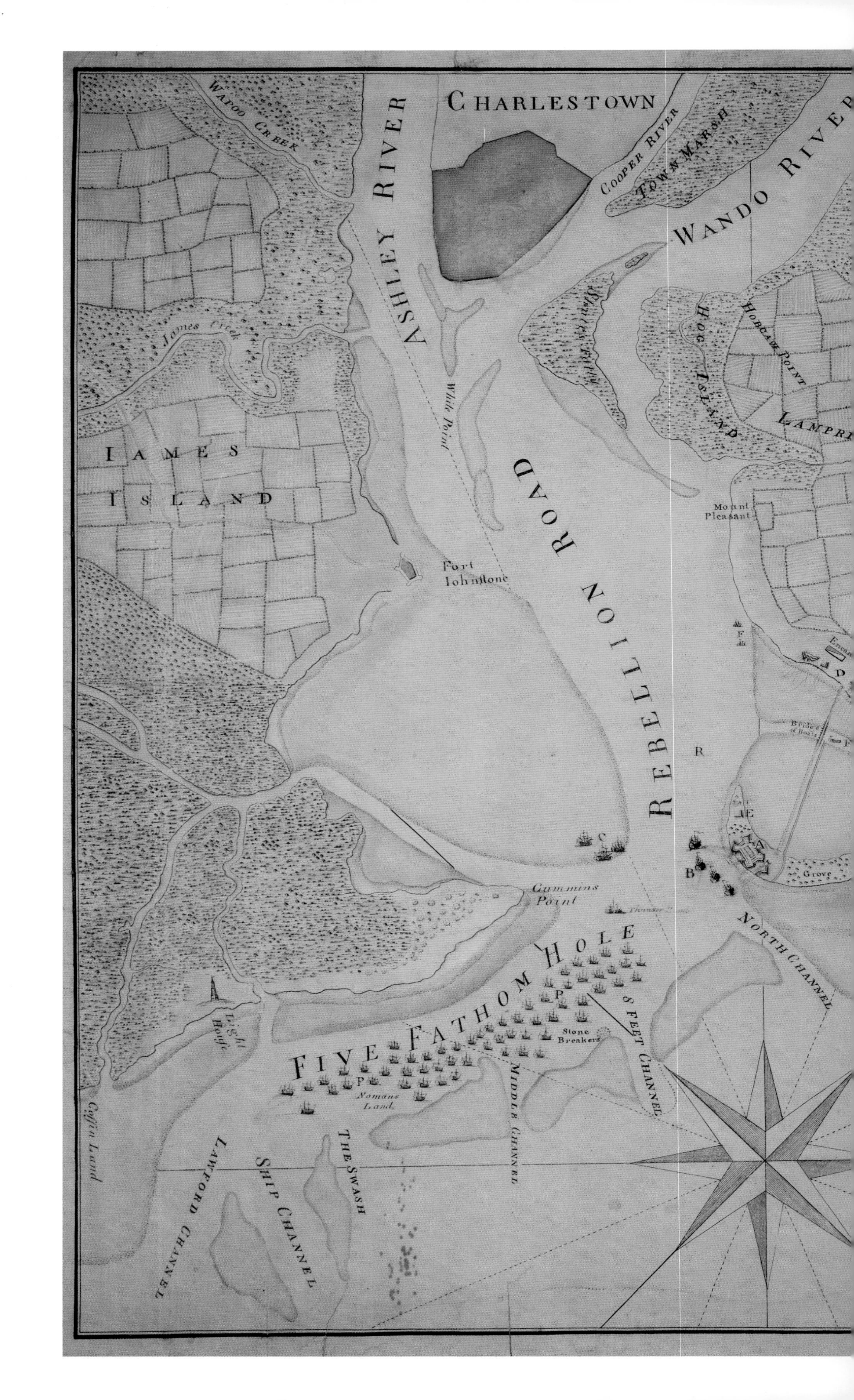

Figure 39. *Plan of the Scene of Action at Charlestown in the Province of South Carolina the 28th June 1776*, John Campbell, manuscript, 60 × 93 cm. Clements Library, University of Michigan

Plan of the SCENE OF ACTION at CHARLESTOWN IN THE PROVINCE OF SOUTH CAROLINA the 28th. JUNE 1776.

REFERRENCE.

A Fort Sullivan. B The Cannonade of Four Ships. C Three Others which got aground early in the day of Action these were intented to have been Stationed at R in order to enfilade the works of the Fort & to Cut off the Communication between Hedrals Point & Sulivans Island. D Hedrals Point & Batteries. E Redoubt. F Rebel Floats & Scooners to Cover the Communication & Bridge of Boats G Encampment of the Army under Genl. Clinton Consisting of about 3000 men. H Channels Fordable at low Water. I Channel once thought to be Fordable & where the Army was to have Crossed to Sullivans Island but proving too deep together with the want of a Sufficient number of Flat Boats to have effected a Landing Otherwise prevented the Army from Cooperating. K A Battery of Small Ordnance Overflown at high Water. L Another Battery to which they Occasionally removed their Guns to. M A Battery of Heavy Ordnance. N Another Battery to which they were to retire to. O Disposition of Armed Vessels & Flat Boats. P Armed Vessels & Transports

John Campbell

LLIVANS ISLAND

Rebel Tents

7 Feet Channel

Green Island

LONG ISLAND

SPENCE'S INLET

DEVISE'S ISLAND

HE ATLANTICK OCEAN

Scale

5 4 3 2 1 0 1 2 3 4 Miles

Clinton departed from New York on January 20, 1776, with two companies of light infantry. (The narrative of this campaign is reproduced in Clinton's memoir *The American Rebellion*.)

The southern campaign was compromised from the start by bad weather and logistical snafus. A planned rendezvous with Commodore Sir Peter Parker's fleet, sailing from Ireland with twenty-five hundred troops, was delayed for over a month. Clinton became pessimistic about his prospects, writing, "The sultry, unhealthy season approaching us with hasty strides, when all thoughts of military operation in the Carolinas must be given up." It was not until June 1 that the combined forces finally reached the coast off Charleston.

The approach to Charleston was precarious. The British engineer John Campbell's manuscript map (figure 39) delineates the shallow shoals that were constantly shifting after tides and storms. Larger ships could enter the harbor through the "Ship Channel" that is identified at the lower left of the map. Parker's flagship *Bristol* and the *Experiment* were too heavy for the channel and their fifty cannon had to be removed for the ships to pass over a sandbar and secure an anchorage in Five Fathom Hole (P).

The waterway to Charleston was called "Rebellion Road," but before entering it Sir Peter's warships had to pass the western tip of Sullivan's Island. Colonel William Moultrie, a grizzled veteran of the Cherokee Indian wars, had hastily erected a fortification there (A). That fort was constructed of palmetto trees that "were brought to us by the blacks, in large rafts," recalled Captain Peter Horry. It was "an immense pen, two hundred feet long and sixteen feet wide, filled with sand to stop the shot." The exiled governor William Campbell had had experience with the Charleston harbor and advised Parker and Clinton against a direct assault from the Wando River as the frontage had been heavily fortified. Instead, he advised attacking Sullivan's Island, "a post of last consequence." Once secured, Charleston would be cut off from supplies and could be easily taken by the British. Campbell volunteered to direct a cannon battery aboard the *Bristol*.

On the morning of June 28 "flattering" winds finally blew in, and Parker was ready for action. Clinton had prepared for this by landing a ground force on adjacent Long Island (G) and he intended to march his troops across the "breach" (H), overwhelm the American positions (M, N), and proceed to the fort. But "to our unspeakable mortification," Clinton explained, "we discovered that the passage across the channel . . . was nowhere shallower at low water than seven feet instead of eighteen inches, which was the depth reported." Instead of marching, they would have to rely on their few flat bottom boats for any assault.

Parker was in command of nine men-of-war ships boasting more than three hundred cannon. Little resistance could be expected from the ramshackle fort. General Lee, having arrived on the scene as American commander, thought the fort "could not hold out half an hour, and that the platform was a slaughtering stage." Moultrie insisted on defending it even though American engineers had failed to complete the bridge (F) that would allow the inevitable retreat. Lee was in the process of removing Moultrie from command when Parker's ships opened fire. He positioned his four most heavily armed vessels close to the fort (B) while sending three frigates (C) around the island's bend to establish a more advantageous position. All of the frigates ran aground on an uncharted sandbar. Years later, Fort Sumter, site of the opening salvo of the American Civil War, was erected on this shallow bar.

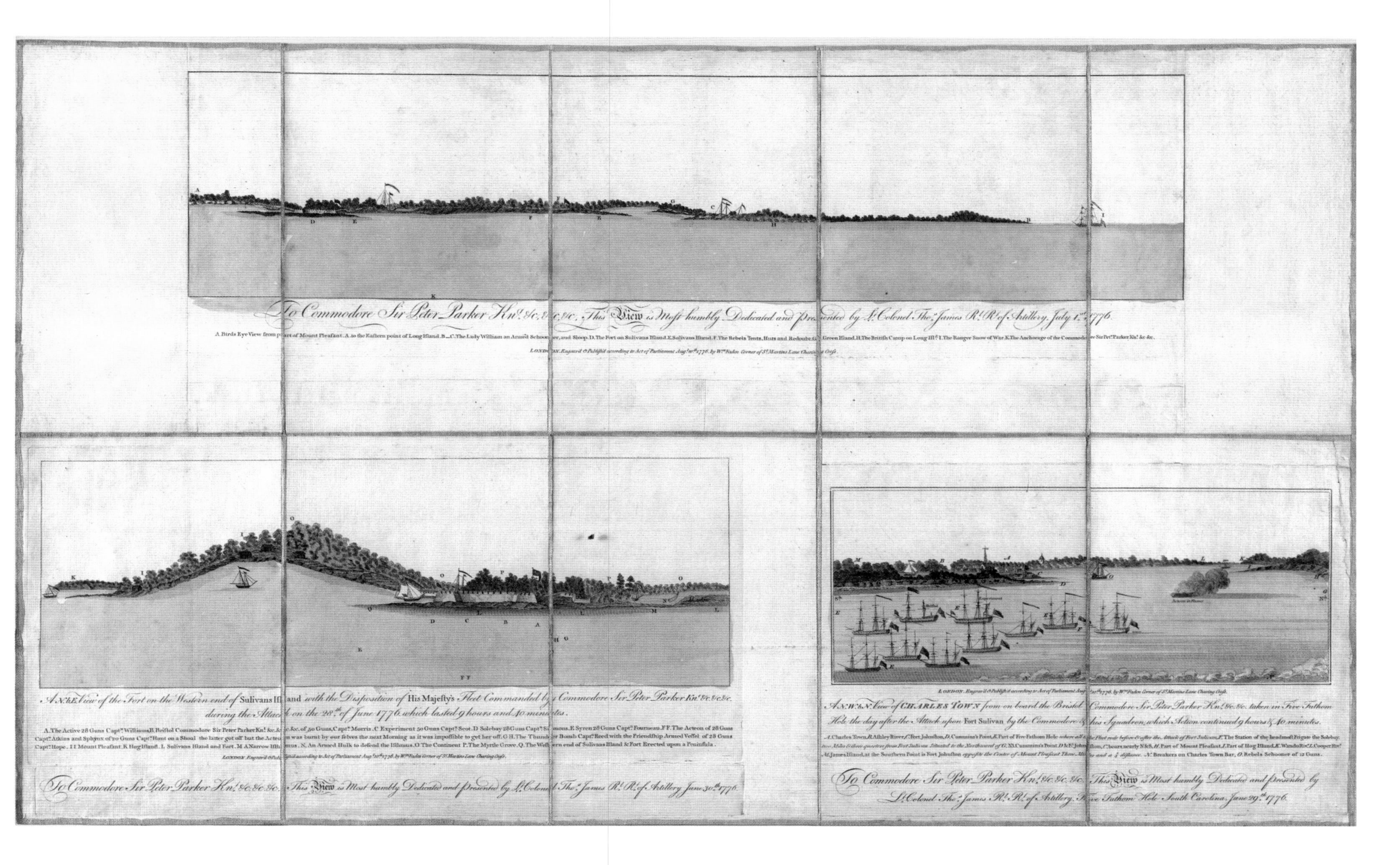

The fire from Parker's warships would have devastated an ordinary fort but the spongy palmetto logs were an excellent defense against the cannonballs. Instead of demolishing the walls of the fort, the cannonballs were absorbed by the logs. Meanwhile, the guns within the fort were remarkably well directed. Moultrie wrote in his journal, "The words that passed along the plat-form by officers and men, were, 'mind the Commodore, mind the two fifty gun ships.'"

Farther off shore, Campbell's map shows the bomb vessel *Thunder.* Artillery Lieutenant Colonel Thomas James accurately aimed the ship's mortars but the shells made little impact. The fort "had a Morass in the middle, that swallowed them up instantly, and those that fell in the sand in and about the fort, were immediately buried, so that very few of them bursted amongst us." Following the battle, Colonel James produced three separate views of the day's events, each subsequently engraved in London by William Faden. An exceptionally rare colored printing of all three formerly in the collection of Michel Capitaine du Chesnoy, Lafayette's personal mapmaker, is shown in figure 40.

The cannonade between Parker's ships and the fort raged for over nine hours. Clinton attempted to ferry his men across the breech, but they were soundly repulsed. By nightfall the flagship *Bristol* lay in tatters—over seventy cannon shells had penetrated her hull. If

Figure 40. *Three Views of the Action at Charlestown, South Carolina*, Thomas James, London, 1776, copperplate engravings, 17 × 58 cm, 20 × 42 cm, 20 × 30 cm. Private collection

powder had not run short at the fort, "the men-of-war must have struck their colors, or they would certainly have been sunk, because they could not retreat, as the wind and tide were against them." Casualties on board the two big ships were exceptionally heavy. Both captains were mortally wounded, as was Governor Campbell. Parker had refused to go below and an exploding shell "ruined Sir Peter's breeches . . . quite torn off, his backside laid bare, his thigh and knee wounded." Andrew Forrester, a crewman on an armed schooner, wrote his brother, "It was impossible for any set of men to sustain so destructive a fire as the Americans poured in upon them."

It was an embarrassing British defeat, especially for Clinton who never engaged his army in the battle. General Howe ordered the forces north to join the New York campaign. On Sullivan's Island, the "pen" was renamed Fort Moultrie to honor its commander. The palmetto tree would join the crescent on the South Carolina flag and inspire a nickname: the "Palmetto State."

New York

On December 7, 2013, a manuscript *Plan of New York Island* (figure 41) was offered for sale at auction in New York City. This map was signed by one of the most accomplished topographical engineers in the British military, Charles Blaskowitz, and it delineated the key events of a battle so ambitious that it was not rivaled until World War I. Four hundred and twenty-seven ships sailed into the New York harbor in August 1776 to put a decisive end to the rebellion. Scholars of the American Revolution took note of this auction lot because no other map of the battle showed the action in and around Manhattan Island in such detail or on such a large scale.

The map had been made expressly for a British Revolutionary War general, Sir William Erskine. Sir William was in New York while much of the action on the map was taking place, and, like his fellow nobleman Lord Percy, the map remained at his estate. Manuscript maps such as those belonging to Erskine, Percy, General Lafayette, and others have been turning up in recent years among the personal papers of the participants in the events. These historically important maps were created as keepsakes of their owners' heroism and were not intended to circulate. Some of the most intriguing recent discoveries of Revolutionary War maps are these personal copies. Nearly a quarter of a millennium after the events, Blaskowitz's *Plan of New York Island* had remained in the hands of the general's family, until Mrs. M. Sharpe Erskine's Trust offered it for sale. The map was completely unknown at the time; it had never been cited in the literature of the war and had never been reproduced in any book or article. Its first appearance in print was in the Christie's catalog.

After Bunker Hill, the humiliated but victorious British had regrouped at their headquarters in Halifax, Canada. For two months, they nursed their wounds and pride while

Figure 41. *A Plan of New York Island, and part of Long Island, with the circumjacent country*, Charles Blaskowitz, 1777, manuscript, 143 × 112 cm. Private collection

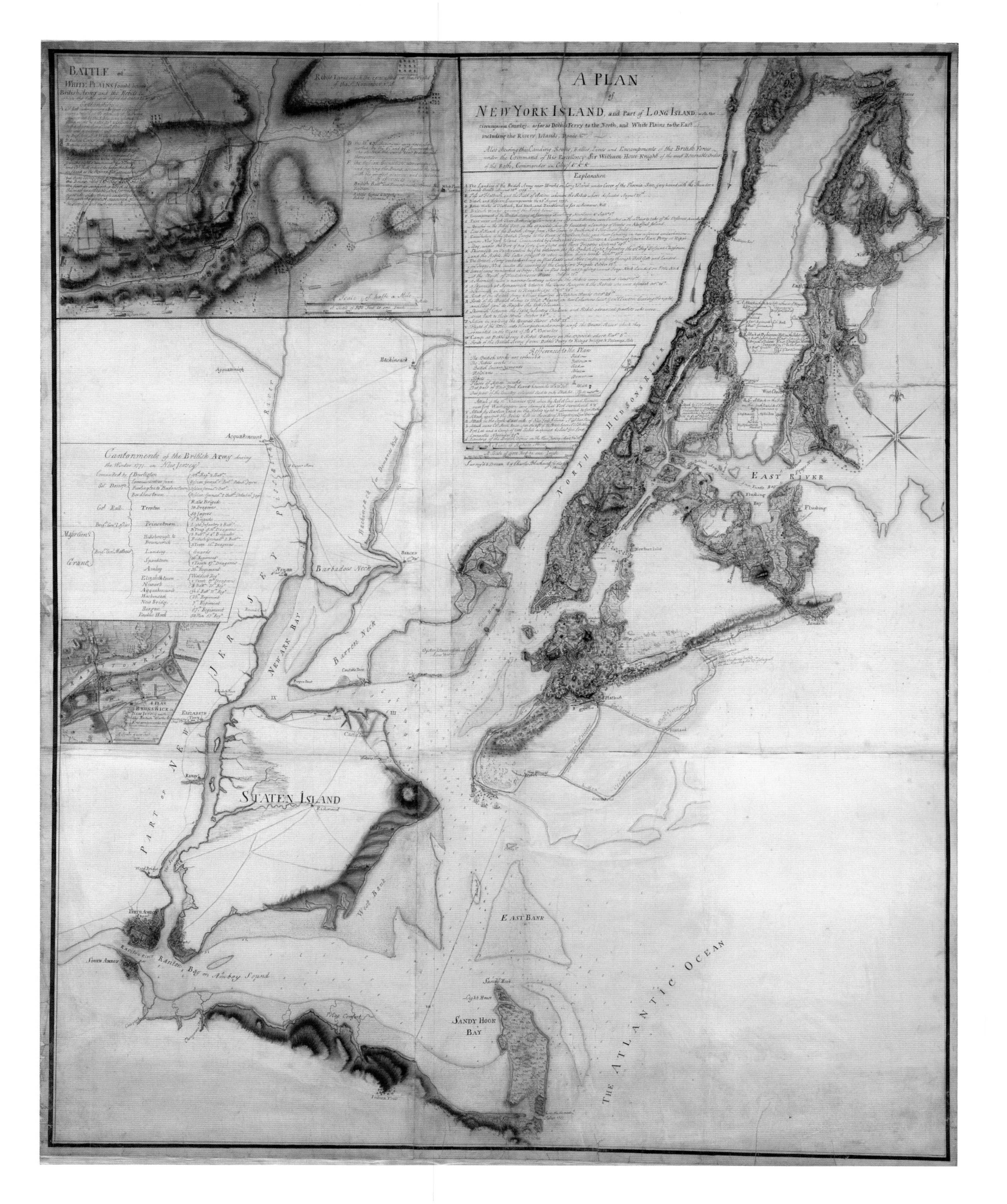

A PLAN of NEW YORK ISLAND, and Part of LONG ISLAND, with the circumjacent Country as far as Dobbs Ferry to the North, and White Plains to the East including the Rivers, Islands, Roads &c.
Also shewing the Landing, Routes, Battles, Lines and Encampments of the British Forces under the Command of His Excellency Sir William Howe Knight of the most Honorable Order of the Bath, Commander in Chief &c. &c. &c.
Explanation
References to the Plan
BATTLE of WHITE PLAINS
Cantonments of the British Army during the Winter 1776 in New Jersey
NORTH or HUDSONS RIVER
EAST RIVER
Flushing
NEW JERSEY
PART of NEW JERSEY
Barbadoes Neck
Barren Neck
NEWARK BAY
Hackinsack
STATEN ISLAND
Richmond
Perth Amboy
South Amboy
Raritan Bay or Amboy Sound
West Bank
EAST BANK
SANDY HOOK BAY
Sandy Hook
Light House
THE ATLANTIC OCEAN
Flatbush
Jamaica

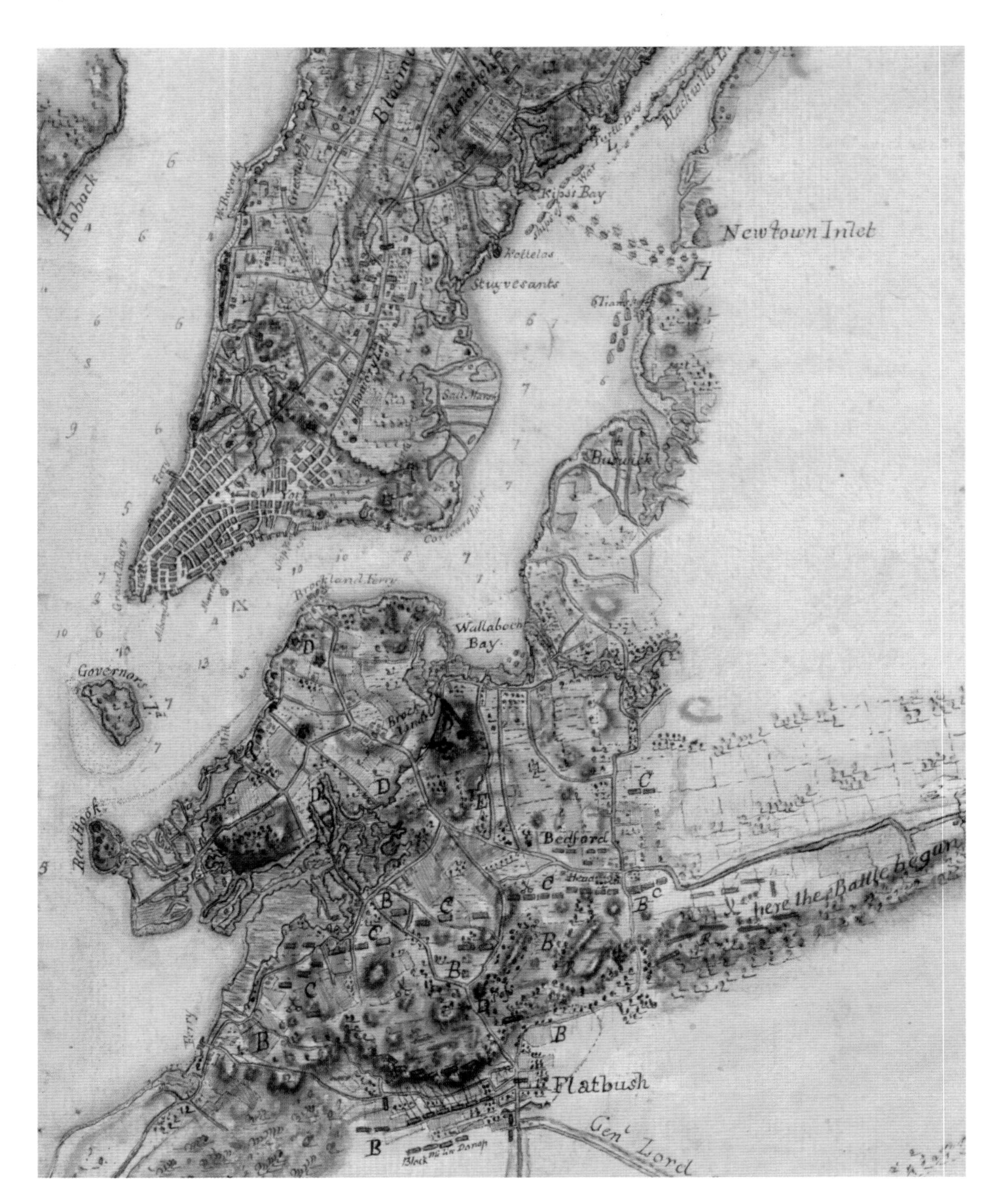

Detail, figure 41, p. 83

hatching plans for a triumph that would reassert their king's authority over his colonies. New York had been the obvious target even before the evacuation of Boston, a strategy anticipated by General George Washington. He had arrived in New York in mid-April and began a troop buildup. New batteries were established and forts were constructed in an effort to thwart British ships from entering the Hudson. But Fort Washington and Fort Lee proved ineffectual when the combined might of His Majesty's army and navy sailed blithely into the New York harbor under the joint command of the Howe brothers, Sir William leading the land forces and Lord Richard the naval. It was the largest show of force the most powerful nation on earth had ever mounted. The Americans had no naval force at all.

On August 22, 1776, the armada crossed the narrows between Staten Island and Long Island transporting thirty-two thousand British and Hessian soldiers and ten thousand seamen to Flatbush and Flatlands in Brooklyn. "The landing of the British Army near Utrecht on Long Island," reads the first line in the legend on the map, "under cover of the Phoenix,

Rose, Greyhound . . . August 22, 1776" (detail, figure 41, opposite). The first encounters of the New York campaign were fought in Brooklyn. The rebels under the command of General Washington had been fortifying this area since February 1776. "By invading Long Island, driving the rebels out of their fortifications, and then threatening New York City with artillery from Brooklyn Heights," wrote the scholar Barnet Schecter, "Howe intended to accomplish his new objectives with a minimum of casualties." The forty-four-gun *Phoenix* and the twenty-eight-gun *Rose* were two imposing British warships in the fleet, and on July 12 Admiral Howe had intimidated Washington and countless others when he flaunted the behemoths on the Hudson River under full sail. A few guns from shore fired anemically at the ships. The cannonballs that the ships returned crashed into houses and careened down city streets filled with panicked residents. For more than a month, these two warships lay at anchor in the Tappan Zee where they controlled all of the communications along the great river.

General Howe landed fifteen thousand troops at Gravesend Bay on Long Island and a smaller force to the north of the bay. More troops, including five thousand Hessians, were added to the force on August 23 and 24. As can be seen on the map, the carefully placed British forces, twenty-one-thousand strong, attacked the poorly organized and ill-equipped rebels at three points on August 27. "Pass at Flatbush and Field of Action," reads the legend on the Blaskowitz map, "wherein the Rebels were defeated August 27." At Flatbush Pass the five thousand Hessians attacked eight hundred Americans under General Sullivan while General Howe's light infantry attacked from the opposite side. Two American generals were captured along with thirteen hundred soldiers who were detained as prisoners. "It was a fine sight," wrote one British officer about all of the American soldiers who had died, "to see with what alacrity the Hessians dispatched the Rebels with their bayonets after we had surrounded them so they could not resist."

The outcome of the first pitched battle of the Revolutionary War was determined in about four hours, and it was a complete rout of the American forces. The surviving American army retreated to Brooklyn Heights where they appeared to be trapped. A fog hovered over the night of August 29 providing fortuitous cover for the troops stranded after the battle. Washington ordered his quartermaster "to impress every kind of craft on either side of New York that had oars or sails" and have them in the East River. Like a moorhen leading its chicks across the swift current of a stream, Washington ferried his surviving troops across the river back to Manhattan. The general's "flawless evacuation of Brooklyn is one of the greatest moments in the annals of warfare," wrote Schecter in *The Battle for New York*. General Howe, on the other hand, has been admonished for not keeping the pressure on the rebels and allowing many of them to escape certain death.

Charles Blaskowitz, the cartographer of these incidents of warfare, was a professionally trained surveyor whose maps of the Revolution, both published and in manuscript, have left some of the best records of American waterways. His survey of Newport, Rhode Island, for example, was engraved by William Faden for inclusion in the *North American Atlas*. Blaskowitz served for more than a decade under Samuel Holland, and the British could not have employed a better mapmaker to chronicle the stages in their decisive victory in New York.

Figure 42. *Plan of the City of New-York as it was when his Majesty's Forces took Possession of it in 1776*, Claude Joseph Sauthier, 1776, manuscript, 80 × 60 cm. Collection of the Duke of Northumberland, Alnwick Castle

Two weeks would pass before Howe would direct his troops on the island of Manhattan on September 15. The explanation on the map reads: "Embarkation of the British Troops under the . . . General Earl Percy." The offensive scattered the weak force of Americans trying to defend the Kips Bay area. Howe's forces easily entered the city where the general later celebrated another victory by sitting down to cakes and Madeira at the home of Robert Murray in Murray Hill. A legend has evolved that during this interlude of refreshment an American force of thirty-five hundred troops, trapped in lower Manhattan, were led by Lieutenant Aaron Burr to the safety of Harlem Heights.

The British exercised their control of Manhattan for only six days before one of the most calamitous events ever to befall the island occurred. A fire broke out on the morning of September 21, 1776, in a wooden house near Whitehall Slip. Some say the cause was arson, while others believe it to have been started in a bordello preferred by British soldiers. A stiff wind was blowing from the south-southeast, disseminating the flames between Broadway and the Hudson River, the densest part of the town. Because most of the houses were constructed of wood and all had cedar shingles the fire made short shrift of the city. Much of it was obliterated, and by the end of the day a third of the city lay in ruins, including 493 houses, many built during the early days of Dutch occupation.

The manuscript map by Claude Joseph Sauthier pictured in figure 42, showing the devastation in striking detail, is among those in Lord Percy's castle in Alnwick, the recently discovered *Plan of the City of New-York as it was when his Majesty's Forces took Possession of it in 1776*. The map scholar William Cumming considered this "Probably the most important example of maps developed from the Ratzer Plan," the standard delineation of the city at the time. It shows, in addition to the area ravaged by fire, forts, embrasures, and ramparts built for Washington to defend the city. In some ways the fire was a blessing for the Americans. Knowing that the British had designs on the buildings for billeting troops, the Americans had thought of intentionally burning the city, but this proposal was summarily rejected by the Continental Congress. "Providence," wrote Washington about the destruction, "or some good honest fellow, has done more for us than we were disposed to do for ourselves."

Blaskowitz and Sauthier were two of the most accomplished military mapmakers of the Revolutionary War and these manuscript maps were among their great achievements. Blaskowitz's map was completely unknown until the twenty-first century and Sauthier's has been almost equally obscure. Sauthier had trained as an architect and surveyor in his native France. There he also learned to ingratiate himself with prominent men, and this ability advanced his career. He arrived in America before 1768 and was soon working closely with, and traveling alongside, William Tryon, the governor of New York. While living in New York, he came under the influence of Lord Percy, and it was for him that Sauthier made this map of the operations on Manhattan. Lord Percy admired Sauthier's work so much that he appointed him his private secretary and was able to persuade the mapmaker to return to bleak Northumberland with him. At Alnwick, he drafted a number of surveys of the Northumberland estate, and these remain among the maps in the archives of the castle.

"Attack of the 16th November 1776," reads the last legend on the Blaskowitz map, "when the Rebel lines and Redoubts near Fort Washington were stormed and that Fort surrendered." That is the date when the British attacked Fort Washington, in the area that is now called Washington Heights, and took possession of New York. By then, most of the American army had left the city for Westchester. General Washington not only had lost the city, he had also lost four thousand men. The outlook for the war was grim. For the next seven years, until the peace was signed in November 1783, New York would remain in British hands and their occupation would be a festering thorn in Washington's side.

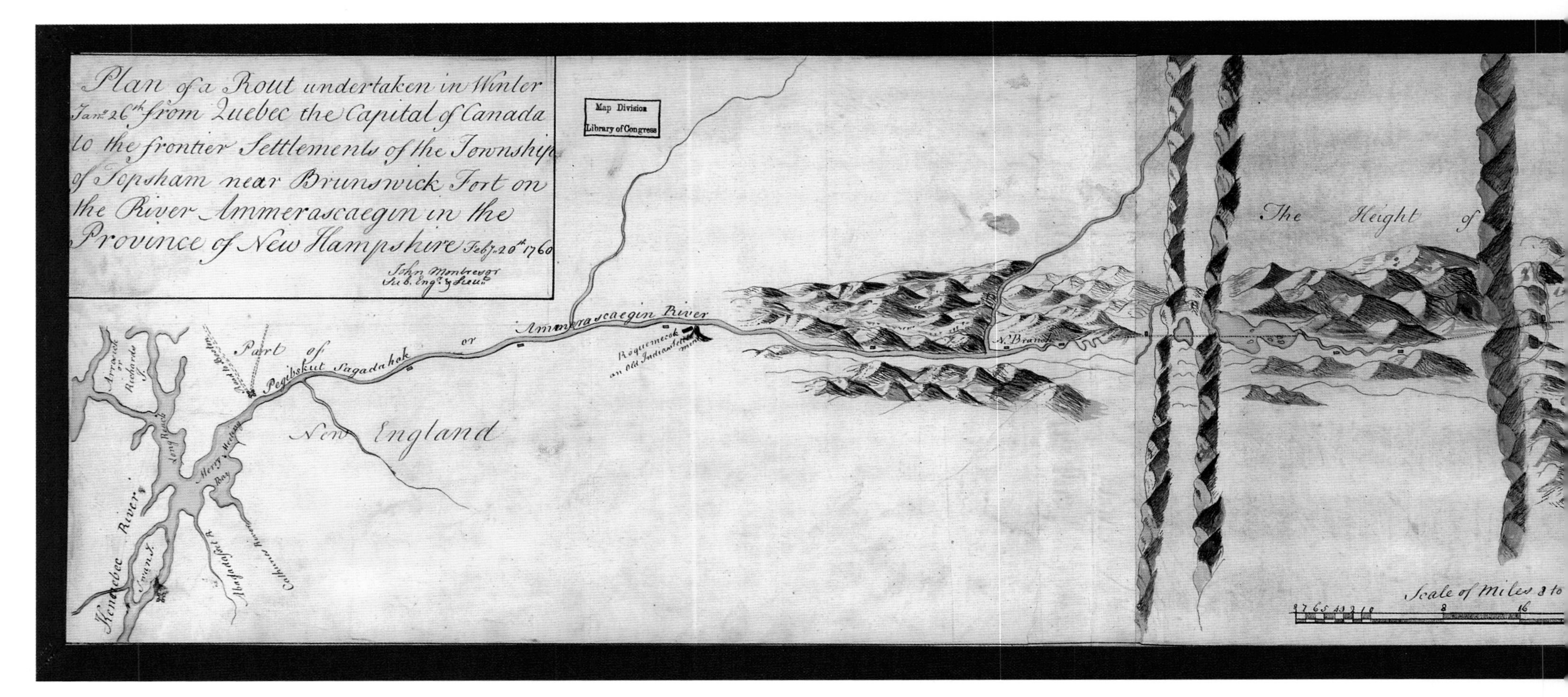

Figure 43. *Plan of a Rout undertaken in Winter, Jany. 26th from Quebec*, John Montresor, 1760, manuscript, 21 × 104 cm. Geography and Map Division, Library of Congress

Quebec 1775

In late January 1760, James Murray, commander of the fortress of Quebec, possessed valuable intelligence for General Jeffrey Amherst, then readying his vast army for a spring offensive against Montreal. Amherst's camp at Crown Point, New York, was only 260 miles south of Quebec but Murray's ignorance of colonial geography led him to send the intelligence by a route many times that distance. He ordered engineer John Montresor to cross the uncharted regions of Maine to coastal New Hampshire, sail to New York City, and then proceed hundreds of miles up the Hudson Valley. Montresor kept a careful record during his military career and, not surprisingly, in his 1760 journal he characterized Murray as "A Madman." On March 23, Montresor finally reached Amherst. In addition to the intelligence, Montresor "Gave the Commander in Chief . . . a plan of my route [through Maine] to scale of 8 miles to an inch." That signed manuscript plan (map) is illustrated in figure 43.

Persuading the colony of Quebec to become the fourteenth state in the Union was high on the agenda of the 1775 Continental Congress. Quebec was larger than all the other colonies combined and her strategic importance was enormous. As long as the British controlled Quebec, the American interior was vulnerable to attacks along the Champlain waterway.

In July 1775 following the Battle of Bunker Hill, George Washington established his headquarters in Cambridge, Massachusetts. A standoff developed with the British troops across the Charles River in Boston, and while Washington considered his options he ordered a force under Philip Schuyler and Richard Montgomery to proceed north from Fort Ticonderoga to capture Quebec. A few weeks later, an ambitious Benedict Arnold, who

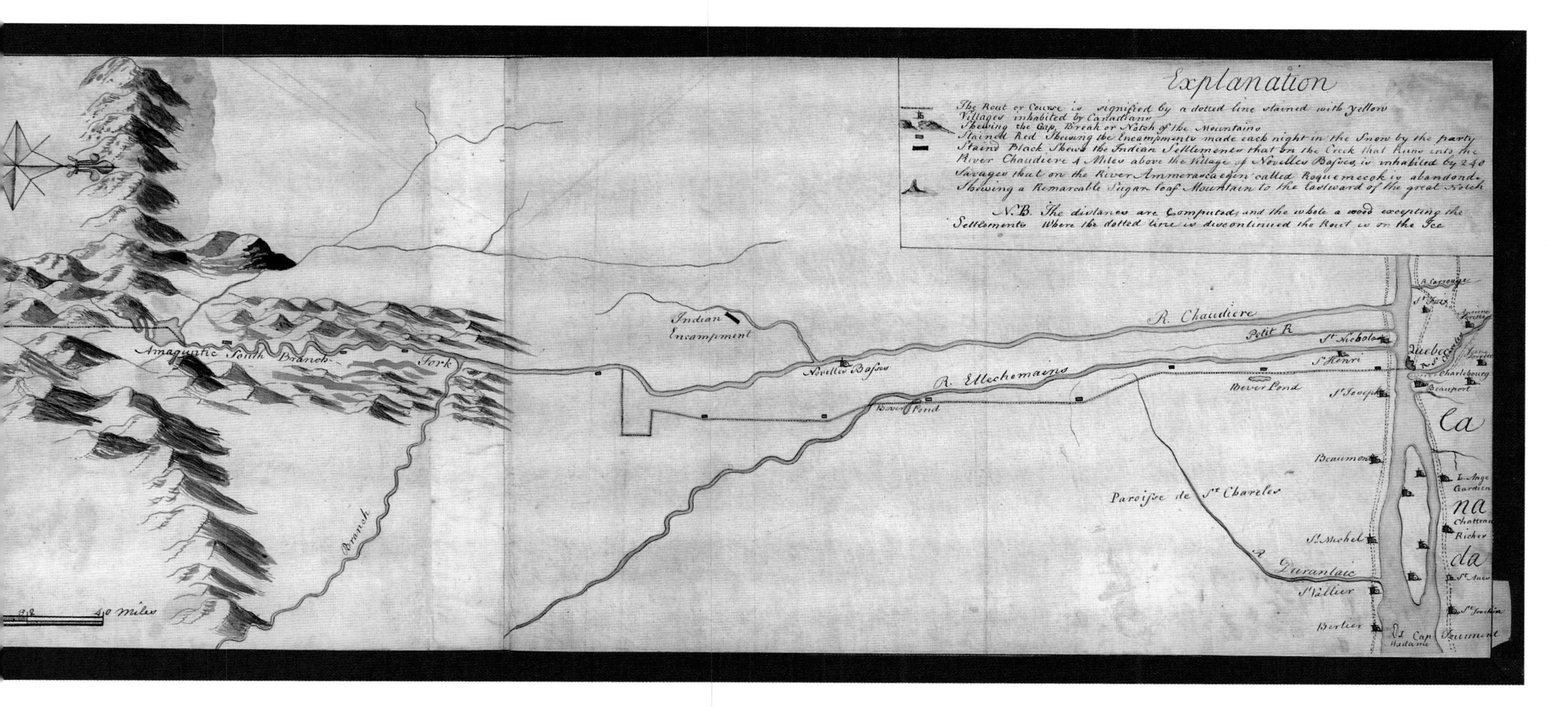

several months earlier played an important role in seizing Ticonderoga, convinced Washington to send a second force to Quebec through Maine. Arnold's "secret weapon" was a copy of John Montresor's journal and map, either the one exhibited here or a similar one from a reconnaissance trip made during the summer of 1761.

Based on the map, Arnold concluded that the journey would be principally by river and he ordered a number of shallow draft bateaux prepared in advance. Three weeks behind schedule, on September 25, the thousand-man army departed from the head of the Kennebec River. In his baggage, Arnold carried several thousand copies of Washington's "Address to the Inhabitants of Canada." Printed in French, the speech passionately encouraged Canadians to overthrow the British government and seize the "Blessings of Liberty."

The journey up the Kennebec past Fort Halifax (lower left on the map) was uneventful but reaching the Dead River (Ammerascaegin) required several portages around rapids and falls. The men struggled with the bateaux, which were constructed of green wood and far heavier than planned. Five days into the trip, the consequences of the late start were already apparent as Captain Simeon Thayer recorded, "Last night, our clothes being wet, were frozen a pane of glass thick."

In the mountainous terrain, rivers became unmanageable. "This morning . . . the country round entirely overflowed, so that the course of the river being crooked, could not be discovered . . . our provisions almost exhausted . . . we have but a melancholy prospect before us," Arnold wrote on October 22. Several days later, in a letter to Washington, Arnold acknowledged the shortcomings of Montresor's map, noting, "I have been much deceived in every account of our route, which is longer and has been attended with a thousand difficulties I never apprehended." Among the difficulties were the bateaux, which shredded on the rapids and lost precious provisions. Arnold's advance force survived on shoe leather and candle wax while waiting to be resupplied by Colonel Roger Enos who commanded the rear guard. But Enos decided it was suicidal to continue forward and he promptly marched back to Massachusetts with almost half of Arnold's army and a majority of the remaining provisions.

The Dead River ends at the rugged Appalachian mountain range separating Maine from Quebec. In order to survive the crossing, the weakened soldiers abandoned the remaining bateaux and, upon reaching Lake Mégantic, the source of the Chaudière River, they were forced to walk the seventy miles to Quebec. There were no paths, and "Barefoot, and the weather Cold and Snow, many of our men died," recorded Henry Dearborn.

As the emaciated survivors reached the first French villages, "The people looked at us with amazement; and seemed to doubt whether or not we were human beings," wrote twenty-one-year-old rifleman George Morrison. Nevertheless, the Québécois treated them warmly and provided provisions. There was every reason to be hopeful that all these people would be on their side against the British.

Figure 44. *Plan of the City and Environs of Quebec, with Its Siege and Blockade by the Americans from the 8th of December, 1775 to the 13th of May, 1776*, cartographer unknown, 1776, manuscript, 46 × 63 cm. Geography and Map Division, Library of Congress

Detail, figure 44, opposite

On November 8, a month behind schedule, Arnold arrived at the east bank of the St. Lawrence River. His army had survived an "impossible" journey only to face the fortress of Quebec with five hundred soldiers and not a single cannon. The manuscript map (figure 44) provides a contemporary depiction of Quebec and the upcoming military actions. The map was subsequently engraved and printed by William Faden, royal geographer to King George III. Faden published a number of battle maps of the American Revolution to satisfy the popular demand for war news in Britain.

Undaunted by the high odds against him, Arnold acquired bateaux from the Québécois and was able to move his army across the St. Lawrence at night and encamp near the fortress. Favorable news soon arrived from Montreal that General Montgomery had taken that town along with a thousand prisoners. The British governor Guy Carleton had narrowly escaped after disguising himself in peasant garb. If Arnold had attacked immediately, the Americans might well have been victorious as the forces within the fort were disorganized and demoralized. But the usually bold Arnold became cautious. A council of war was called and it was generally agreed to wait for the arrival of Montgomery's forces and cannon.

On November 19 Carleton dramatically returned to Quebec and was shocked by what he saw. "We have so many enemies within . . . I think our fate extremely doubtful," he wrote to Lord Dartmouth the next day. The Québécois were terrified of Arnold's army. According to the British captain Thomas Ainslie, "The report spread that these people were insensible of cold & wore nothing but linen in the most severe seasons–the French word *toile* (linen) was changed into *tolle* (iron plate) and the rumor then ran that the Bostonois were musket proof, being all cover'd with sheet iron."

In early December, Montgomery arrived with six hundred soldiers and, senior to Arnold, assumed overall command. "He is tall and very well made; and possesses a captivating address . . . His recent successes give us the highest confidence in him," noted Morrison. Montgomery had been a British career officer before he moved to America in 1773, married into the wealthy Livingston family, and changed his allegiance to the patriot cause.

Montgomery hoped to draw the British troops from the fortress but Carleton would not budge. The snow was three feet deep, smallpox and other diseases raged through the camps, and many enlistments were set to expire on December 31. Montgomery's only alternative was to storm the fortress, but in a final letter to his brother-in-law Robert Livingston he acknowledged such an action defied all the odds of siege warfare.

Montgomery prepared a night attack on the fortress but bright moonlight made this impossible until a storm developed on December 31. Montgomery and Arnold were to attack from opposite ends of the lower town (L and M on the Faden map detail, page 91). Once that objective was secure, they planned to attack the upper town with scaling ladders. Carleton had established barricades in the lower town but Montgomery and Arnold took their first obstacles easily. Then, as recorded by the American officer Francis Nichols, "A drunken sailor swore he would fire one shot before he would retreat, went to a gun loaded with grape shot, and with a match fired it off, and unfortunately for us killed the brave Montgomery . . . Col. Campbell who usurped the command (for his rank was quartermaster) ordered a retreat. If Col. Campbell had advanced and joined Col. Arnold's troops, he would have met with little opposition." At the other end of town, a musket ball shattered Arnold's leg and he was carried from the battle. Captain Daniel Morgan continued to lead the attack for several hours, but without the aid of Montgomery's troops he was eventually surrounded and forced to surrender.

In his tent, Arnold held two cocked pistols anticipating a last stand but Carleton did not press his advantage. He would let the American army suffer through the harsh winter outside Quebec's secure gates and wait for reinforcements when the ice cleared. For several months, Arnold stubbornly maintained the trappings of a siege, setting up a few cannon across the St. Charles and St. Lawrence Rivers as indicated on Faden's map. In late April Arnold left for Montreal after General John Thomas arrived with reinforcements and assumed overall command. Thomas had scarcely settled in when British ships, loaded with troops and supplies, arrived on May 6. The next day Carleton marched a sizable force out of the fortress gates and the Americans began a disorganized and panicked retreat south.

Valcour Island

"I am on my way to Canada," Benjamin Franklin wrote Josiah Quincy on April 15, 1776, "detained by the present state of the lakes. . . . I have undertaken a fatigue that, at my time of life, may prove too much for me, so I sit down to write a few friends, by way of farewell." Delayed by frozen water, the septuagenarian emissary was on a mission, sanctioned by the Continental Congress, to persuade Canada to join the Union. Franklin would live

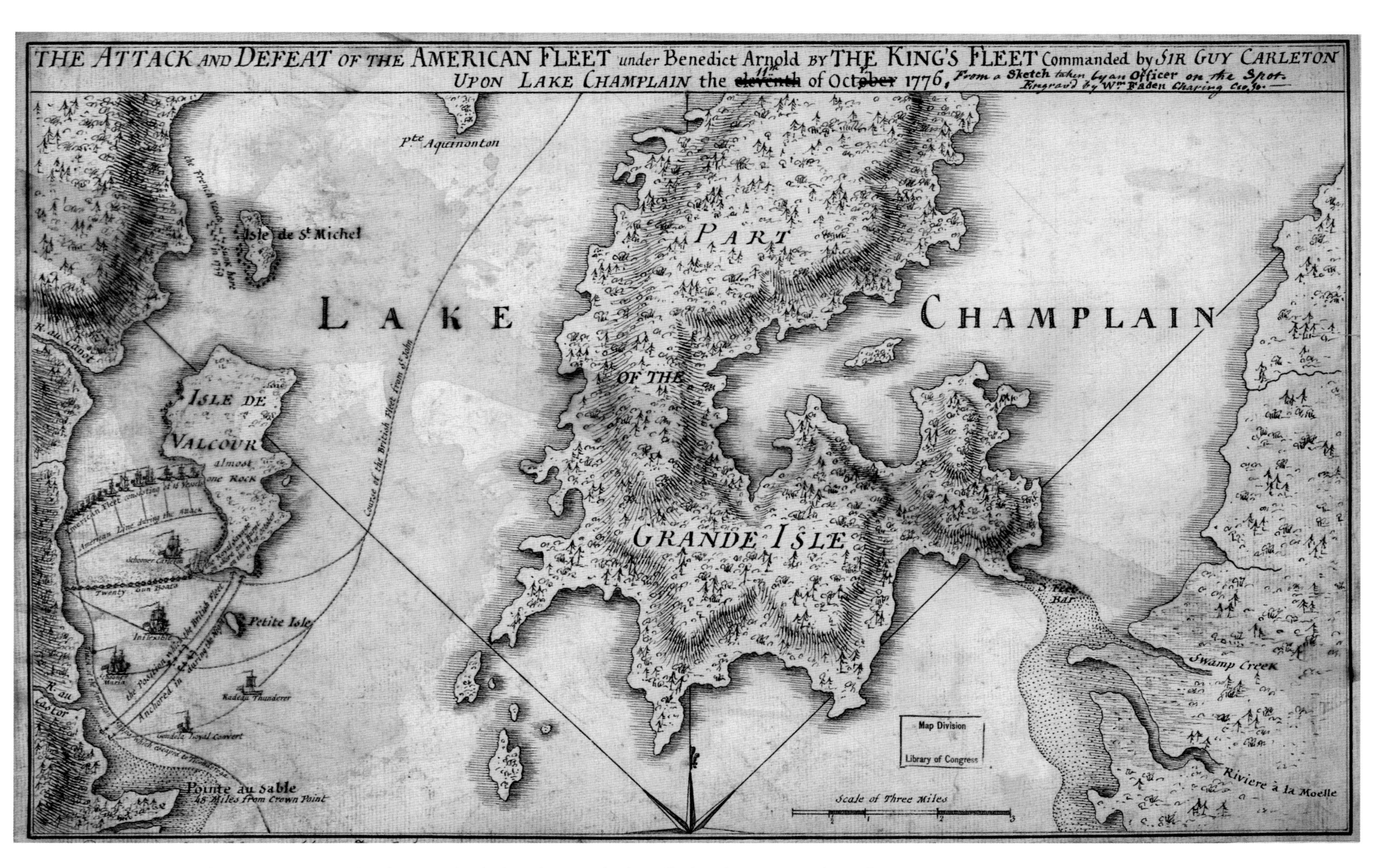

another fourteen years but hopes of a deal with Canada dwindled within two weeks. After discussions with General Benedict Arnold and Canadian representatives in Montreal, the Continental Army abandoned Quebec. Franklin left town on May 11 convinced that "it would be easier to buy Canada than to conquer it."

After commanding general John Thomas died of smallpox, Benedict Arnold was left in charge of the beleaguered army in Canada. Arnold was the last American to leave Montreal, overseeing the retreat to Fort Ticonderoga on Lake Champlain. This set up a major conflict on the lake with General Guy Carleton. In 1776 several armed vessels that had been captured by the Americans the previous year controlled the water, but more ships were needed in order to make a strong showing against the British who were in pursuit. The preparations for battle consisted mostly of shipbuilding. Arnold recruited hundreds of shipwrights to build workshops on the lake and construct a fleet of warships. In Quebec, Carleton was also building ships but his were much larger, too large, in fact, to navigate the Richelieu River to reach Lake Champlain. They had to be built in sections and assembled on the lake. Building boats is time-consuming business, and it was not until October 4 that the British launched the flagship *Inflexible* to join twenty-four other British vessels to oppose Arnold's fifteen vessels, mostly galleys and gondolas.

Figure 45. *The Attack and Defeat of the American Fleet under Benedict Arnold by the King's fleet commanded by Sir Guy Carleton upon Lake Champlain the 11th of Octr. 1776*, unknown mapmaker, 1776, manuscript, 25 × 42 cm. Geography and Map Division, Library of Congress

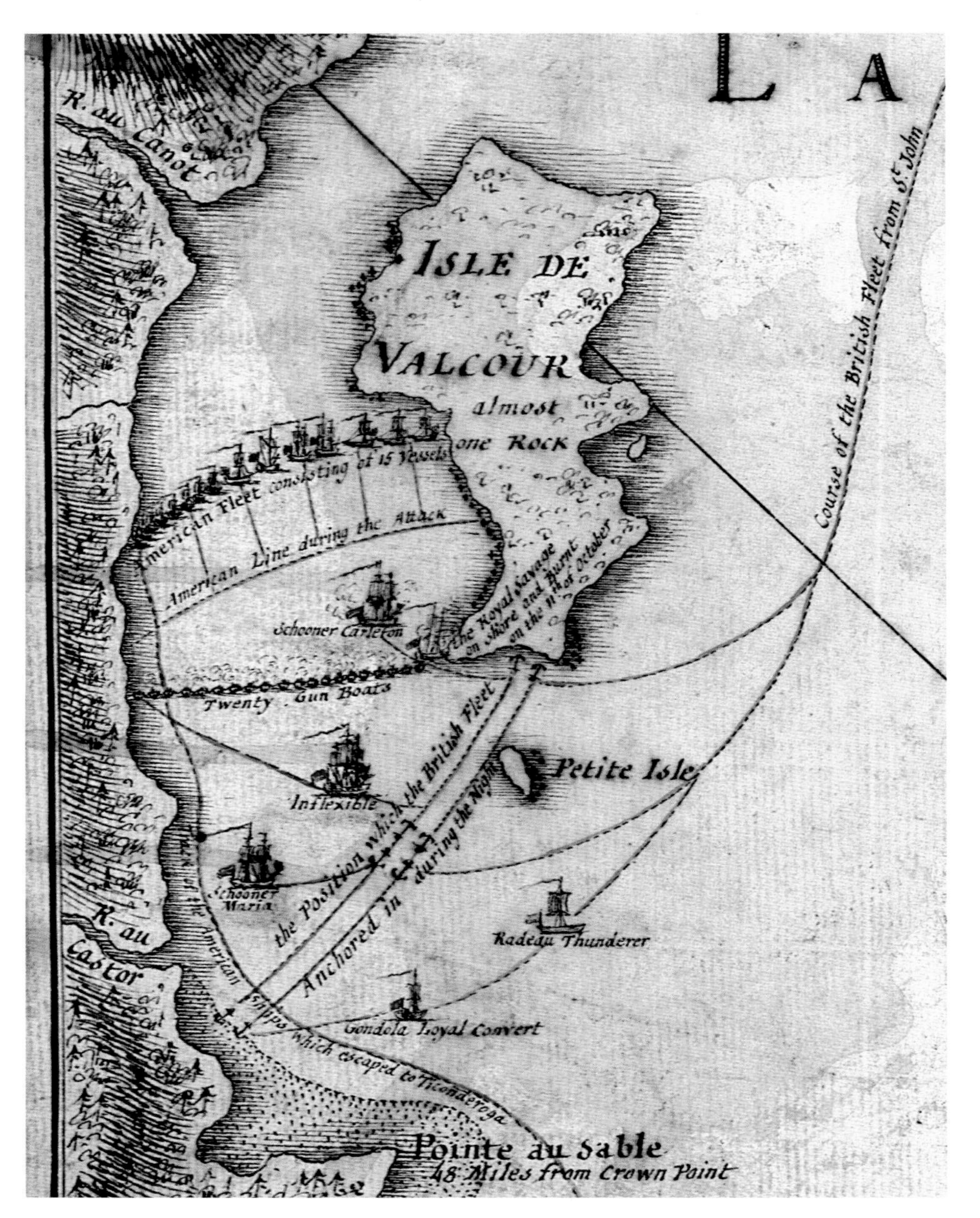

Detail, figure 45, p. 93

The Attack and Defeat of the American Fleet under Benedict Arnold is a manuscript map "taken by an officer on the spot" (figure 45). William Faden published a later version of the map in London on October 11, 1776. Arnold had planned to engage the enemy on the open water, but when he learned that several of the British ships carried more firepower than his entire fleet, he sought a location where he might at least have a fighting chance. He settled on Isle de Valcour (Valcour Island), located on Lake Champlain roughly seventy miles north of Fort Ticonderoga, and featured in the detail (left).

Valcour is described as "almost one rock," whose peak above the waterline exceeded one hundred feet. This height and its westward peninsula allowed Arnold to hide until the British sailed well past the island and lost the power of the wind. The "American Fleet Consisting of 15 Vessels" was situated "as near together as possible," wrote Arnold, "and in such form that few vessels can attack us at the same time, and those will be exposed to the fire of the whole fleet."

The windy conditions hampered the British mariners, and they were able to maneuver only the *Carleton* and the oared gunboats into fighting positions. The fierce exchange of fire lasted for several hours, and when dusk descended over the lake the *Carleton* was disabled, a number of American boats sustained heavy damage, and many officers and crew had been killed. Aboard the battleship *Congress,* Arnold fought ferociously, aiming most of the cannon himself, but he knew that the next morning his weakened fleet would be surrounded. A less tenacious commander might have surrendered, but Arnold and his officers sprang into action. Overnight he moved his fleet out of danger "with so much secrecy that we went through them entirely undiscovered." The next morning, General Carleton could scarcely contain his rage. Daring American escapes had become commonplace during the Revolution, diminishing victories of the overconfident British. The map traces the "Track of the American Ships which escaped to Ticonderoga."

The British found the hobbled American fleet but Arnold escaped again. "The sails, rigging, and hull of the *Congress* were shattered and torn in pieces," wrote Arnold, "the First Lieutenant and three men killed, when, to prevent her falling into the enemy's hands, who had seven sail around me, I ran her ashore in a small creek . . . set her on fire with four gondolas, with whose crews, I reached Crown-Point through the woods that evening, and very luckily escaped the savages, who waylaid the road in two hours after we passed."

The British claimed victory on the lake: "by straining every nerve in their country's cause, we have outdone them [the rebels] in working, as much as in fighting." History would judge the result differently. As A. T. Mahan wrote, "The little American navy on Champlain was wiped out; but never had any force, big or small, lived to better purpose or died more gloriously, for it had saved the Lake for that year." At Fort Ticonderoga, twelve thousand militia had joined the remnants of the retreating American army to buttress its defenses. A twenty-year-old officer, John Trumbull, who would later immortalize the American Revolution in his paintings, wrote in his autobiography, "Our appearance was indeed so formidable, and the season so far advanced (late in October), that the enemy withdrew without making an attack."

The battle on the lake caused a reshuffling of personnel. Benedict Arnold's bold tactics had delayed the British advance and his status as a military leader reached mythical proportions. General Burgoyne returned to London where he criticized Carleton for not following up on the naval victory and successfully lobbied the king for the following year's command.

Fort Ticonderoga

"I hear Burgoyne has kicked Ticonderoga into one of the lakes–I don't know which, I am no geographer," wrote Horace Walpole. He was not the only Englishman lacking knowledge of North American geography. Ticonderoga, tucked into the continent's wild interior, was terra incognita to most of his fellow countrymen, including the ministers in London planning war strategy. King George III could consult his trove of North American maps, but their accuracy did not extend far beyond the coastlines. It fell to a Frenchman to produce the most important map of Ticonderoga during the Revolutionary War.

When the nineteen-year-old Gilbert du Motier, Marquis de Lafayette, arrived in America on June 13, 1777, he brought with him one of the most skillful mapmakers of the era: Michel Capitaine du Chesnoy. As aide-de-camp, Capitaine would serve at Lafayette's side and record the young general's battles over the next four years. The masterful maps he drafted of these encounters are an important record of Lafayette's military career and include several of the most significant maps of the American Revolution. Despite their historical importance, Capitaine's maps were produced almost exclusively in manuscript form and are therefore exceedingly rare.

The manuscript map exhibited here, *Plan of Carillion or Ticonderoga* (figure 46), shows the position of the British forces at Ticonderoga, New York, on October 24, 1777. Capitaine usually drew battlefield maps after he had made a firsthand reconnaissance. However, in October 1777 he was still recuperating in North Carolina from an illness contracted on his voyage from France, and Lafayette himself was nursing a wound received while serving under General Washington at Philadelphia. Capitaine must have compiled his map of the fort from sources now lost, for his delineation is unique–not copied from any known map, published or unpublished. The detail provided on British and Hessian troop positions several months after their victory at Ticonderoga and several days after their defeat at Saratoga suggests that Capitaine gained access to enemy documents. Unlike Capitaine's other maps the legends on this one are in English rather than French.

The French built Fort Carillion (Ticonderoga) in 1756 on a promontory overlooking the confluence of Lake Champlain and the inlet to Lake George. Cannon mounted atop the fort's star-shaped bastions controlled the passage into the American and Canadian interiors and the location had been bitterly contested during the French and Indian War. In 1775 Americans captured Ticonderoga from a small British garrison but barely retained the fort in 1776 when Carleton withdrew his army at the end of the campaign season.

In November 1776, when both sides went into winter quarters, only about two thousand soldiers were left at Ticonderoga, and General Horatio Gates appointed Captain Anthony Wayne as commander. Known as "Mad Anthony" for his combustible personality and aggressive military tactics, Wayne was commended by Gates to the Continental Congress as having "the health and strength fit to encounter the inclemency of that cold, inhospitable region." On December 4, 1776, the durable commander provided the Pennsylvania Council of Safety with a description of winter quarters: "The wretched condition the battalions are now in . . . is shocking to humanity and beggars all description–We have neither beds nor bedding for our sick to lay on or under . . . no medicine or regimen suitable for them; the dead and the dying lying mingled together in our hospital or rather house of carnage."

While American generals planned the 1777 campaign, little was done to improve the defenses or manpower at Ticonderoga. General Arthur St. Clair assumed command in June 1776 and found insufficient troops to man the outlying redoubts, the fort's first line of defense. General Gates's chief engineer, Thaddeus Kosciuszko, was also at the fort. He immediately recommended the construction of a battery on the heights of nearby Sugar Loaf Mountain across from Fort Ticonderoga, as shown on Capitaine's map. St. Clair refused to take his advice. Kosciuszko was also concerned that an enemy fleet could easily surround Ticonderoga, leaving only a retreat through the wilderness to the north. To mitigate this he ordered defenses at Mount Independence across the lake to the south, connected to Ticonderoga by a log and heavy chain pontoon bridge that served to move both troops and supplies and as a barrier to enemy ships. Both are shown on Capitaine's map.

By the end of June General Burgoyne was advancing on the fort with a force of eight thousand. As Kosciuszko had feared, Burgoyne's first order of business was to place a battery atop Sugar Loaf, leaving both Ticonderoga and Mount Independence at risk of bombardment. St. Clair held a council of war, and without firing a shot in defense he called for a retreat. The advance guard of Burgoyne's force nipped at the heels of the retreating Americans but the enterprising Kosciuszko covered the withdrawal by felling trees in the trail, building dams, and destroying bridges.

The loss of Ticonderoga was viewed as one of the American army's most devastating defeats. Upon hearing the news George III exclaimed, "I have beat them! I have beat all the Americans!" George Washington had never seen Ticonderoga but he had adopted the widespread view that the place was impregnable. In a letter to General Schuyler, a typically understated Washington described the withdrawal as "not within the compass of my reasoning: this stroke is severe indeed and distresses us much." John Adams went so far as to suggest that "we must shoot a general before we can win a victory." For leaving Ticonderoga without mounting a defense St. Clair was court-martialed, although he was eventually acquitted. After Burgoyne fell at Saratoga, St. Clair's defense–"We lost a post but saved a province"–gained a small measure of acceptance.

Figure 46. *Plan of Carillon ou [sic] Ticonderoga: which was quitted by the Americaines in the night from the 5th to the 6th of July 1777*, Michel Capitaine du Chesnoy, 1777, manuscript, 68 × 52 cm. Geography and Map Division, Library of Congress

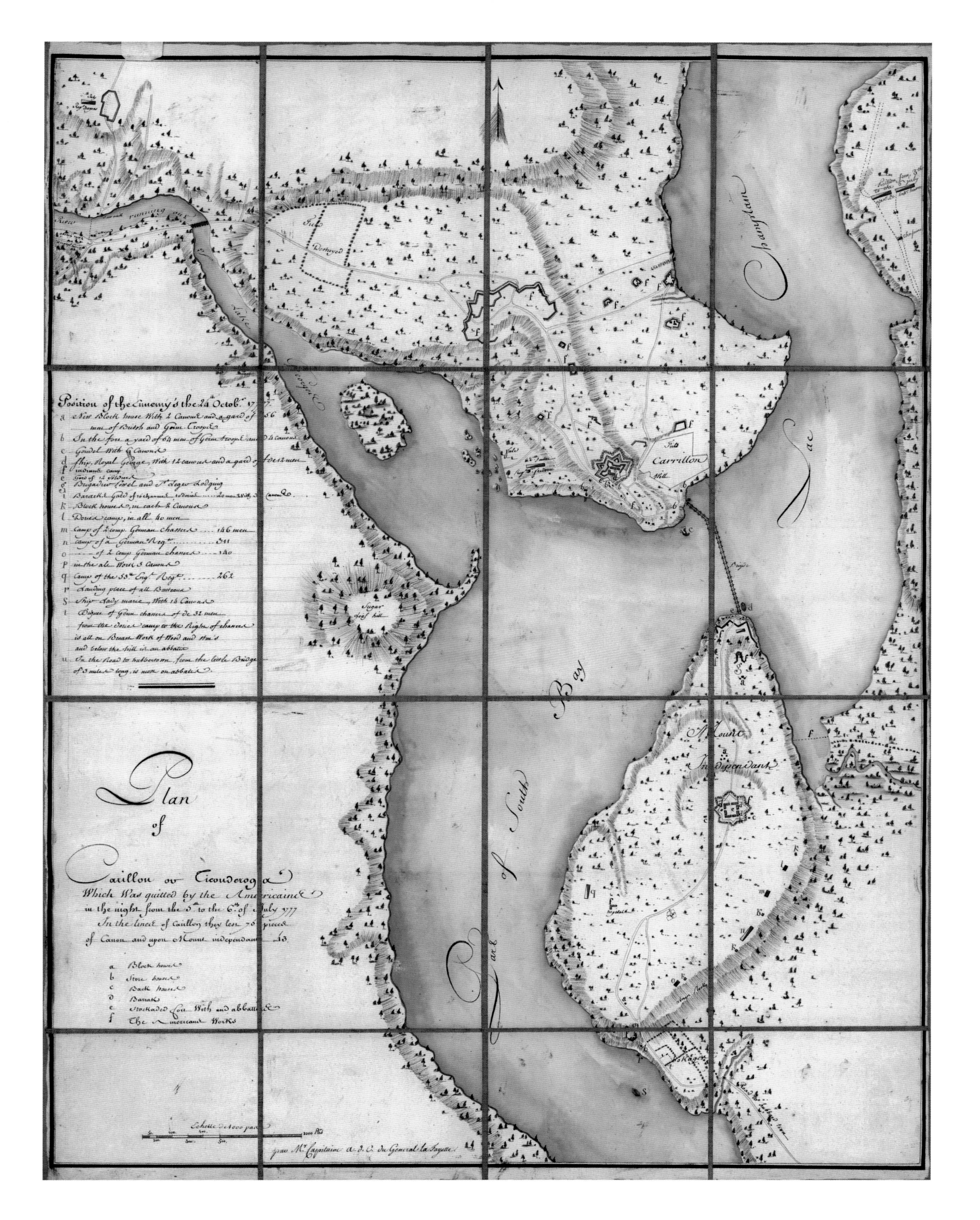
Position of the Enemy's the 24th Octob. 1777
a New Block house With 2 Canons and a gard of 56 men of Britsh and Germ. Troops
b In the fort a gard of 64 men of Germ troops and 4 Canons
c Goudel With 6 Canons
d Ship, Royal George, With 12 canons and a gard of de 12 men
f indians camp
e Gard of 12 soldiers
g Brigadier Powel and St Leger Lodging
i Baracks Gard of 16 chasseurs, 10 dories ... 20 men with 3 Canons
k Block houses, in each 2 Canons
l Dories camp, in all 40 men
m Camp of 2 comp. German chassers ... 146 men
n camp of a German Regt ... 311
o ... of 2 comp German chassers ... 140
p in the ale Work 3 Canons
q Camp of the 53d Engl Regt ... 262
r Landing place of all Batteaux
s Ship Lady marie, With 14 Canons
t Piquet of Germ chassers of de 32 men
from the dories camp to the Right of chassers is all on Breast Work of Wood and Ston's and below the hill is an abbatis
u In the Road to hubbertown, from the little Bridge of 3 miles long, is more on abbatis
Plan of Carillon or Ticonderoga
Which Was quitted by the Americains in the night from the 5th to the 6th of July 1777
In the lines of Carillon they left 76 pieces of Canon and upon Mount independant 43
a Block houses
b Store houses
c Back houses
d Baraks
e Stockaded fort With and abbatis
f The Americans Works
Carrillon
Mount Independant
Sugar loaf hill
Bay
South
Champlain
Lac
Destroyed
hospital
Echelle de 1000 pas
par Mr Capitaine A. D. C. du General la Fayette

Trenton

November 16, 1776, "brought us Hessians glory," Andreas Wiederholdt recorded in his diary. General Wilhelm von Knyphausen had led a charge up the rocky slopes to Fort Washington, the American stronghold overlooking the Hudson River. When the Hessians reached the fortress walls, twenty-eight hundred Americans surrendered as word had spread of savage bayonet charges that granted no quarter to wounded and retreating soldiers. Across the river in Fort Lee, New Jersey, George Washington broke down and wept. He had witnessed the entire battle and knew these soldiers were headed to prison ships in Brooklyn's Wallabout Bay where they would probably die of disease and starvation. Only one man in four survived the ordeal.

The Hessians, who had outfought the Americans at the Battles of Long Island and White Plains, seemed invincible. It would have been inconceivable to Wiederholdt that five weeks later, on the banks of the Delaware River, a pivotal battle of the Revolutionary War would be lost by the Hessians, and he would record it in his diary and on the manuscript map illustrated in figure 47.

Hesse-Kassel was a small and poor German principality whose largest export was soldiers. George II had employed Hessians to fight in the Seven Years' War, and Andreas Wiederholdt had distinguished himself in Europe during the war, moving up through the ranks to become a lieutenant in 1760, but the peace that followed stalled his advancement. When George III required soldiers in America, Hessians signed on in large numbers, including Wiederholdt. In all, 30,000 German "mercenaries," most from Hesse-Kassel, would serve in the British army during the Revolution; 23,000 would survive and 4,800 of those remained in America.

In the wake of the siege of Fort Washington, the fort was renamed Knyphausen, and Colonel Johann Rall was decorated for gallant leadership. His brigade had killed more rebels than any other unit in the British army. What was left of Washington's army retreated through New Jersey. "The splendid appearance," wrote Major James Wilkinson, "and triumphant march of the British battalions in pursuit of our half naked, sickly, force, overspread the country with terror." Sensing weakness in the ranks, General Howe offered amnesty to anyone who signed an oath of allegiance, and thousands accepted. Such leniency was no way to crush a rebellion, according to Wiederholdt, who described the population's "fear of us Hessians": "They did not believe that we looked like other human beings, but thought that we had a strange language and that we were a raw, wild and barbaric nation." Charles Stedman, a captain under General Charles Cornwallis, described the plunder of the countryside, writing, "The friend and the foe from the hand of rapine shared alike . . . The odium began with the Hessians . . . though the British troops were far from escaping a share of the imputation."

Cornwallis doggedly pursued Washington but was never able to catch him. At the Delaware River, the American general transported his army across in advance of Cornwallis. Stedman wrote sarcastically that the British generals "appeared to have calculated with the

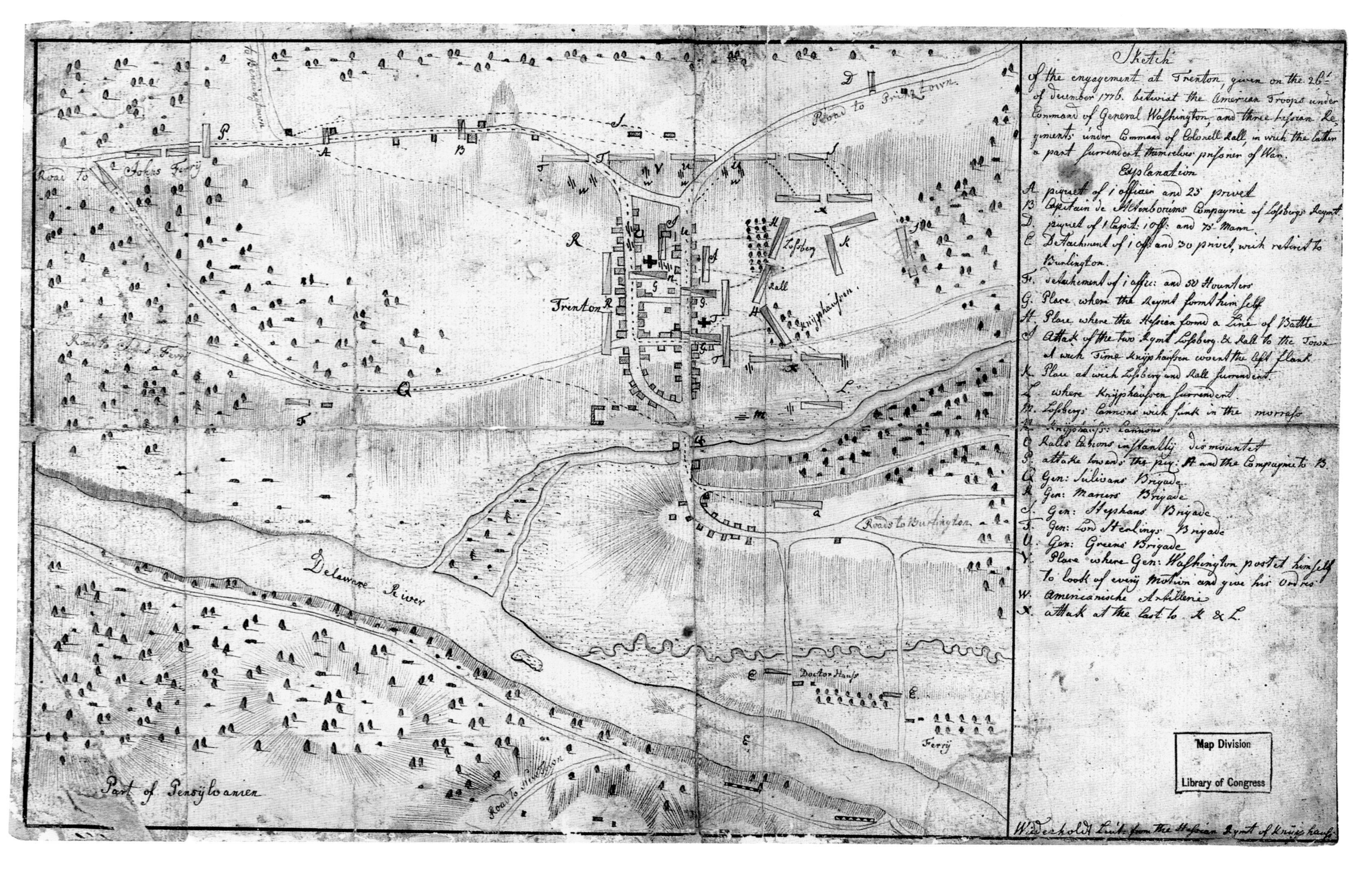

greatest accuracy the exact time necessary for the enemy to make his escape." Worse yet for the British, the Americans had removed all the boats from the New Jersey side of the river. Howe considered waiting a month until the Delaware River froze, but instead of marching over the river to Philadelphia he decided to go into winter quarters. Winter campaigns were not popular in the eighteenth century, and besides, Howe had struck up a romance with the charming wife of his commissary general. Washington's defeat could be deferred to the spring, he reasoned. This proved to be a serious error in judgment.

Figure 47. *Sketch of the Engagement at Trenton, given on the 26th of December 1776*, Andreas Wiederholdt, 1776, manuscript, 23 × 28 cm. Geography and Map Division, Library of Congress

Of the winter cantonments, Trenton, along the banks of the Delaware River, was the most strategic. Three Hessian brigades, fourteen hundred men in all, guarded it with Colonel Rall in command. According to Wiederholdt, Rall was a music lover and while in winter quarters he was more interested in the performances of his fife and drum corps than the protection of the town. A senior officer proposed erecting some defensive positions, but Rall dismissed the suggestion: "Unnecessary nonsense." "Let them come!" Rall said. "Why defenses? We will go at them with the bayonet!"

On the night of December 25, Washington silently crossed the ice-clogged Delaware River nine miles above Trenton. The operation's name, Victory or Death, characterized the

Figure 48. *Henry Knox*, Charles Willson Peale, ca. 1784, oil on canvas, 58 × 48 cm. Courtesy of Independence National Historical Park

desperate state of the American cause. Washington relied on the Massachusetts "Marble-headers" to ferry his twenty-four hundred men and supplies safely to the New Jersey side. Among the troops was Colonel Henry Knox (figure 48), who dragged eighteen cannon along snow-encrusted roads. Cannon were more effective than muskets in wet weather because their barrels and fire holes could be kept dry with rags.

"I formed my detachment into two divisions, one to march by the lower or river road, the other by the upper or Pennington road," wrote Washington, who wanted to surprise the Hessians after they had spent a drunken night celebrating Yuletide. The army was three hours late when it assembled on the Delaware River. "This made me despair of surprising the town as I well knew we could not reach it before the day was fairly broke." The top left portion of Wiederholdt's map shows the "Penningtown" road as well as the river road ("Road to John's Ferry").

Andreas Wiederholdt was in charge of the advance Piquet guard (A) on the morning of December 26 and reported, "About an hour after sunrise . . . I was attacked out of the woods . . . I gave the order to fire and engaged them until I was nearly surrounded by several battalions. I then pulled back under a steady fire, to the Alten–Bouckum Company" (B). A retreat into the town followed, where Wiederholdt reported the events to Rall, just awakened after passing a merry evening at the home of a local loyalist.

Hundreds of Hessians were able to flee across the bridge at the southern end of town before General Sullivan's division (Q) arrived along the river road to oppose them. Rall was far too proud and held the rebels in too much contempt to consider a retreat. He mounted his horse and called to his regiment: "Forward march! Advance! Advance!" The Hessians formed in their traditional lines with fife and drums playing but Knox's cannon (W) opened with devastating effect. Several days later Knox wrote to his wife, "The hurry fright and confusion of the enemy was not unlike that which shall be when the last trump shall sound." Nearby, at (V), "Washington posted himself to look at every motion and give his orders." Rall fell mortally wounded attempting to rally his soldiers. Wiederholdt details the location of the American brigades (yellow) and Hessian brigades (blue). The latter are surrounded and nine hundred men including Wiederholdt had no choice but to surrender.

The impossible had succeeded, and its effect was profound. "It may be doubted whether so small a number of men ever employed so short a space of time with greater and more lasting results upon the history of the world," wrote the historian George Otto Trevelyan about the Battle of Trenton.

The exhausted, half-frozen Americans and their prisoners returned to Pennsylvania. Several days after the battle Washington invited Wiederholdt and several other Hessian officers to dine with him. "He did me the honor of conversing extensively with me . . . As I frankly spoke my opinions, that our dispositions had been bad and otherwise we would not have fallen into his hands, he asked me what better dispositions I would have made." Wiederholdt provided this information in detail and it is also likely he made the attached map at the request of Washington, who personally coveted maps.

Wiederholdt's diary contains a map of the Trenton battle in German, while the map shown here is in English, a language in which neither Wiederholdt nor any of the other Hessians was proficient. The illustrated map also includes the American brigade commanders by name, something Wiederholdt would not have been able to provide without working with an English-speaking collaborator. Interestingly, the map is housed today as a part of the Rochambeau collection at the Library of Congress. Its presence among the other maps seems unusual and it is interesting to speculate that Washington may have given it to the Comte de Rochambeau sometime during their joint campaign leading up to Yorktown in 1781.

Wiederholdt was released in a prisoner exchange in early 1778 and the following year he was en route to Halifax when his ship *Triton* was disabled during a hurricane. After floundering at sea for several days Wiederholdt saw ships on the horizon but "after they approached and displayed their flag with thirteen stripes, our joy turned to sorrow." Andreas Wiederholdt was a prisoner once again.

Philadelphia

The original plans for the siege at Albany called for General William Howe to transport his army from New York harbor, ascend the Hudson River Valley, and then join General John Burgoyne, who was descending from Canada. The American rebels would have been no match for the combined forces of these two armies. General Howe, however, had so much confidence in Burgoyne that, instead of meeting his fellow general in Albany, he and his sixteen thousand troops sailed out of the harbor on July 20 in 245 transports and supply ships guarded by sixteen additional warships, all under the command of his brother, Admiral Richard Howe.

Intelligence of Howe's armada reached the disbelieving George Washington at his headquarters in Morristown, New Jersey. The commander in chief found the exodus from New York nothing short of bewildering. Where were these Howes going? They could have been on their way to the Carolinas or perhaps to New England. Their destination was in fact the Delaware River and then on to Philadelphia, the political nerve center of the rebellion and the seat of the Second Continental Congress. Like Burgoyne, Washington expected Howe to be headed to the Hudson River. "Howe's in a manner abandoning General Burgoyne is so unaccountable a matter," wrote Washington on July 28, "that till I am fully assured it is so, I cannot help casting my eyes continually behind me."

Figure 49. *Portrait of Marquis de Lafayette*, Charles Willson Peale, 1779, oil on canvas. Washington-Custis-Lee Collection, Washington and Lee University, Lexington, Virginia

That month of July 1777, the American commander in chief made the acquaintance of an idealistic teenager who would have a profound effect on the outcome of the war and on the lives of both men. This was the Marquis de Lafayette (figure 49), and to him the moment of his meeting with General Washington was the most electrifying of his life. He had come to America as an independent agent with ambitions to participate in the fight for liberty and become an officer in the American army to boot. Washington appointed him major general, and Lafayette would see action in the Philadelphia campaign.

By the end of August the Howes were well on their way up the Chesapeake Bay and on August 25 they were landing troops at Head of Elk, the northernmost point of the bay, before assembling the forces off the Delaware capes with Philadelphia in their sights. "This was probably the most serious mistake that Howe made during his two years of command in America," so believed the historian R. Ernest Dupuy. This site of debarkation was some fifty-five miles from Philadelphia and delayed his campaign against the city by almost a month. And strategically Philadelphia was a less important stronghold than the British had envisioned.

Washington had a little time to prepare for the coming assault, despite the large number of Continental troops deployed in the area in and around Saratoga. Washington's army numbered around fifteen thousand and was quartered near Wilmington, not far from the Brandywine Creek. Howe's route from the Chesapeake to Philadelphia passed through Chadds Ford on the Brandywine, near Washington's headquarters.

As the British advanced, they encountered no resistance until they reached the Brandywine, about twenty-five miles southwest of the city. There, Washington and Lafayette had established positions. The Battle of Brandywine began at dawn on September 11, 1777, and Lafayette was drawn into combat for the first time in his life when he engaged British troops under the command of General Charles Cornwallis. Employing antiquated military tactics learned in his native France, Lafayette dismounted and boldly led his astonished men in charging the enemy. It did not take long for a musket shot to penetrate his leg below the calf. Nothing could have thrilled the youthful major general more. The wound was not serious, and it provided a palpable initiation into the military. From that time on, Lafayette proudly displayed the scar as an accessory–his first badge of courage.

Major John André, a handsome young officer in the British army, executed one of the most intriguing maps of the battle at the Brandywine Creek (figure 50). Considered one of the most accomplished military men of his age, the popular officer was also a skilled draughtsman and mapmaker. His diary and papers at the Huntington Library in San Marino, California, include numerous manuscript maps in his hand. His *Battle of Brandewyne on the 11th September 1777* indicated in yellow "the places near which the Rebels made most opposition" against the British regulars. General Howe had dispatched Wilhelm von Knyphausen to "begin early to cannonade the Enemy on the opposite side," wrote André, to distract the rebels while Howe and Cornwallis drove their enemies "from wood to wood on every side."

Much of the activity of the battle takes place in the lower left of the map. Chadds Ford is situated along the Brandywine, and the Heights, above the river, are boldly delineated. It was from there that "General Knyphausen cannoned in the morning; they presuming the whole army to be there." Below Chadds Ford is the section of the river where the rebels "crossed to attack part of Gen. Knyphausen's Corps"; the "Hill from which they were repulsed" is on the other side of the river.

The British outnumbered the Americans at Brandywine and had far fewer losses. Between twelve hundred and thirteen hundred rebels were either captured or killed; General Howe lost fewer than six hundred. Major André and others felt that once Howe had the upper hand in the battle, he should have followed up with a total annihilation of Washington's army. Instead he called off his troops, allowing the Americans to retreat and fight another day. After the battle, André surveyed the battlefield and made a disquieting

Figure 50. *Battle of Brandewyne on the 11th September 1777* [in the Journal of John André], 1777, manuscript, 18.5 × 30 cm. Reproduced by permission of the Huntington Library, San Marino, California

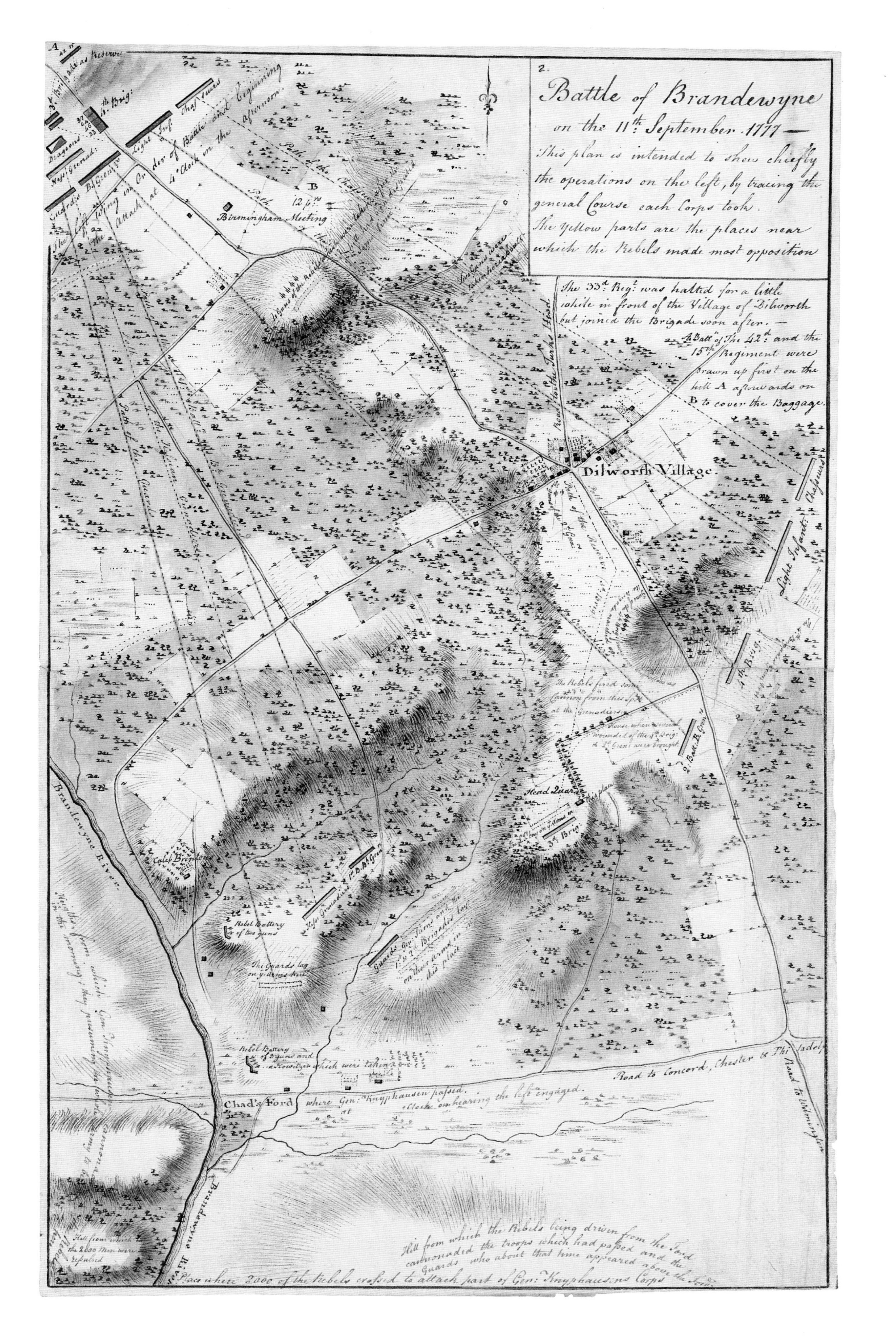
2.
Battle of Brandewyne
on the 11th. September 1777 —
This plan is intended to shew chiefly
the operations on the left, by tracing the
general Course each Corps took.
The Yellow parts are the places near
which the Rebels made most opposition
The 33d. Regt. was halted for a little
while in front of the Village of Dilworth
but joined the Brigade soon after. —
2d. Batt. of The 42d. and the
15th. Regiment were
drawn up first on the
hill A afterwards on
B to cover the Baggage.
Dilworth Village
Birmingham Meeting
Light Infant. Chasseurs
4th. Brig.
3d. Brig.
Head Quar.
Chasseurs
Dragoons
The Rebels fired some
Cannon from this Spot
at the Grenadiers
Rebel Battery
of two guns
The Guards lay
on the Rebels' Hut
Rebel Battery
of 8 guns and
a Howitzer which were taken
Chad's Ford
where Gen: Knyphausen passed.
Road to Concord, Chester & Philadelphia
Road to Wilmington
Brandewyne River
Hill from which the Rebels being driven from the Ford
cannonaded the troops which had passed and the
Guards who about that time appeared above the Ford
Place where 2000 of the Rebels crossed to attack part of Gen: Knyphausen's Corps
Hill from which
the 2000 Men were
repulsed

discovery. Five of the brass guns that he recovered from the defeated enemy had been made in France and were probably brought by Washington's new acolyte Lafayette. In the end, the presence of the French on American battlefields and waters would doom the British effort to maintain colonial rule.

When the Americans were unable to thwart the British advance at Brandywine, Howe continued his march toward Philadelphia. One obstacle in his path was Paoli, Pennsylvania, five miles from his objective. The name of this town had come from Pasquale Paoli, the Corsican general who had fought for that island's independence and was known for his great love of liberty. Washington had ordered General Anthony Wayne to take up a position to harass Howe's troops there, but the British learned about the mission and Wayne's location; Major General Charles Grey was dispatched to surprise the American general with a night attack on September 20. The battle took place after midnight and is noted for the use of swords, bayonets, and a deluge of blood. Major André "heard steel piercing flesh, strangled death cries. In a few minutes, several hundred men who were trying to flee were skewered." American losses numbered two hundred and Wayne was soundly defeated. The "Paoli Massacre," recalled one British participant, was "altogether the most dreadful I ever beheld." From there, the British reached the city of Philadelphia unopposed.

While some troops of Howe's army were stationed in Philadelphia under the command of General Cornwallis, the main force was camped five miles away in Germantown. It was there that on October 4, 1777, the Americans made an audacious attack on the camp. At this juncture, a decisive victory over Howe would have gone far in changing the character of the war. Washington's plan called for a surprise attack on sleeping British troops by converging on them from four roads that entered the town from different directions. Were it not for a number of setbacks the nocturnal strategy might have succeeded. First, there was a breakdown in the coordination of the attacking units. Unexpected delays prevented the troops from arriving on the scene together. Orders became confused and a dense fog engulfed the battlefield. The battle ended in chaos as many Americans died from friendly fire; there were, in fact, heavy losses on both sides. General Howe's dog became so disoriented in the fog that he joined the Americans as they retreated. Washington returned him to his owner. Washington retreated to Valley Forge for the winter and the British took up residence in Philadelphia.

From Philadelphia, Major André wrote to his mother: "I believe I shall be in a very quiet station henceforth. Mr. Washington [André had great contempt for Washington and could not even call him "General"], it is said, has divided his army, and means to seize outposts and attack small detachments. You know, *we generals* have nothing to do with retail fighting." The dashing officer enjoyed the pleasures of Philadelphia, frequenting balls and engaging in romances. Some of his time was taken up with a theatrical group called Howe's Thespians. André and his fellow actors performed thirteen plays. Three years later, on September 23, 1780, after being apprehended in civilian clothes behind American lines, André was hanged as a spy by the Continental Army.

Major André was not the only cartographer at the scene of the Philadelphia campaign. Charles Blaskowitz, one of the greatest mapmakers of the Revolutionary War, crafted a

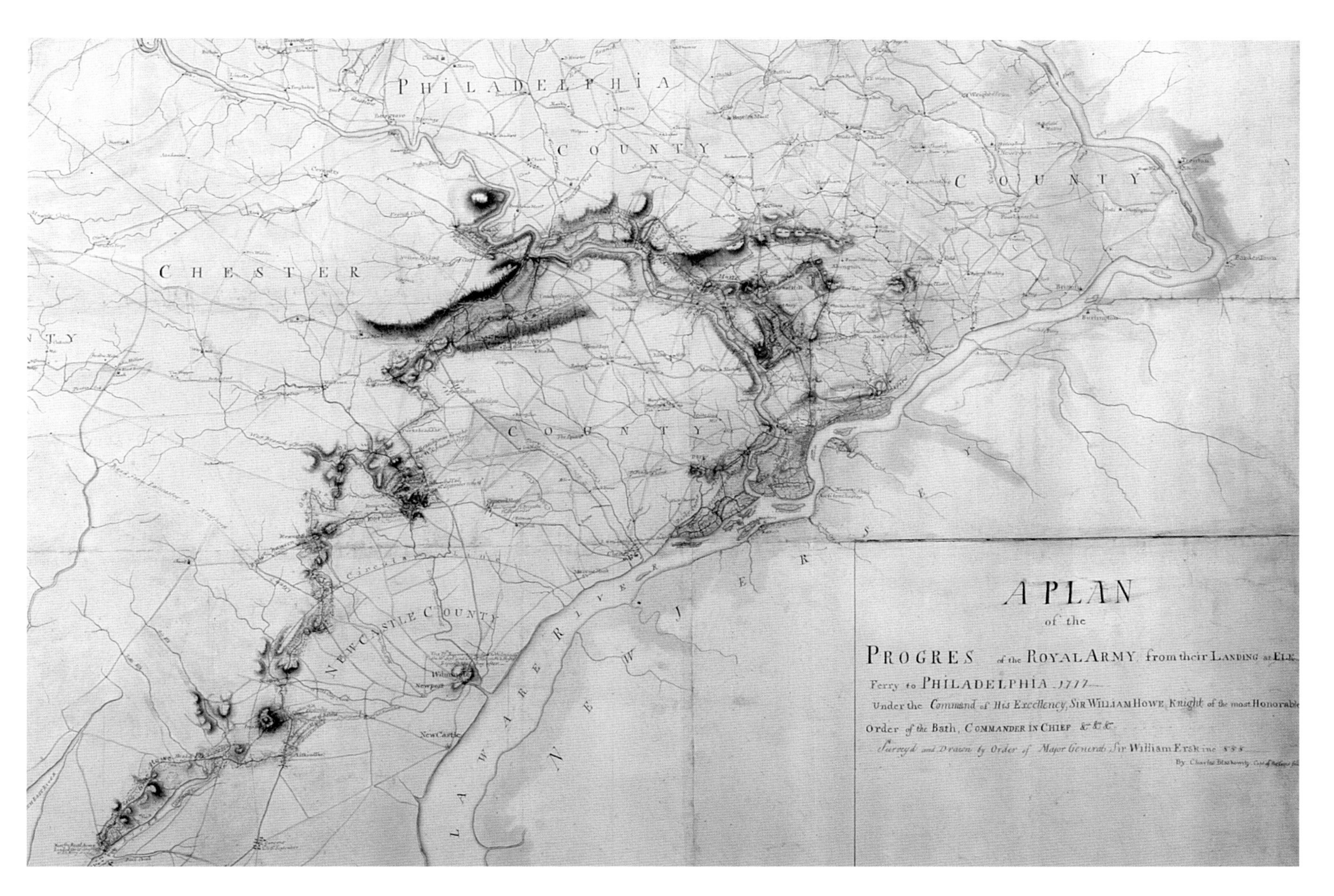

Figure 51. *A Plan of the Progress of the Royal Army from their Landing at Elk Ferry to Philadelphia* (detail), Charles Blaskowitz, 1777, manuscript, 131 × 135 cm. Courtesy of Graham Arader

large and detailed map of "the Progres of the Royal Army from their Landing at Elk Ferry to Philadelphia 1777," a detail of which appears in figure 51. This map was unknown to scholars and collectors before 2010 when out of nowhere it appeared for sale at a Christie's auction in New York City. The map is not dated but Alexander Johnson, who cataloged the map for Christie's, places its execution somewhere between December 22, 1777, the date of the last event depicted, and May 1778, as Sir William Howe is still identified as commander in chief, a post he vacated that month. The map itself chronicles the daily progress of the British army as it moved from the Chesapeake to Philadelphia. Battlefields along the way, such as Brandywine and Germantown, are delineated with precision, and such small locations as taverns and meetinghouses are also included.

Philadelphia was not as important a stronghold as the British expected. Washington's army had weathered the campaign reasonably well, and in its wake his troop numbers increased and they were becoming better trained. What did the capture of Philadelphia actually provide for the British war effort? In fact, little changed after the city fell. When informed of the capture, Benjamin Franklin reportedly remarked, "No, Philadelphia has captured Howe!" Howe resigned his post and the British evacuated the city in June of 1778.

Saratoga

In the mid-nineteenth century, the English historian Edward Creasy published his book *Fifteen Decisive Battles of the World*. Some of the conflicts he wrote about are legendary engagements of antiquity. The Battle of Hastings and the Defeat of the Spanish Armada make the list, as do the achievements of such exalted military leaders as Napoleon and Wellington. One of these decisive battles took place on American soil. During the early years of the Revolutionary War, the British had won a complete victory in New York and were well on their way to shoring up the unconditional domination of the rebellious colonists. All they needed was a single convincing victory to consummate it, and they decided that the best way to achieve this was to attack the colonies along the Hudson River. The rebels had defeated the British in small skirmishes in New Jersey, at Trenton and Princeton, but they had not yet won a conclusive battle over their enemy. This they did at Saratoga, and the triumph was the turning point of the American Revolution.

In 2013, Jim Lacey seconded Creasy's high opinion of the Battle of Saratoga in his *Moments of Battle: The Twenty Clashes That Changed the World*. More than a century and a half had passed since the publication of Creasy's book, and the American Civil War and two world wars had been fought in the interim. There were now twenty decisive battles in the history of the world, and Saratoga still had a prominent ranking. The key player on the British side was General John Burgoyne (figure 52). Known as "Gentleman Johnny," he had observed the battle at Bunker Hill with great alarm and had returned home to England in November 1775 with his mind set on decimating the colonials. In London, he hatched a belligerent plan with Lord George Germain, the colonial secretary. Despite his lofty position in colonial affairs, Germain had never been to America and had no knowledge of the geography of the colonies and little understanding of the tenaciousness of its rebels.

Their plan called for Burgoyne to be given command of the British army in Canada. From Montreal, the general would head south to confront the enemy along the Hudson River. While Burgoyne was on the move, General William Howe, who had made Manhattan his base of operations since his conquest of New York City, would move north, and the two would join forces in Albany as a single, formidable army and take control of this strategically located stretch of the river. Essentially this would isolate the New England colonies and pave the way for the British to divide its subjects into easily conquerable entities.

Little, however, went according to plan. Traveling from Montreal to Albany proved to be far more arduous than it had seemed in the well-appointed planning rooms in London. The difficult terrain caused many delays and hardship, but a more disastrous defect involved General Howe. When Burgoyne arrived in Saratoga, thirty-three miles north of Albany, he learned that Howe was on his way to Philadelphia instead, in pursuit of General Washington's Continental Army. Administering a war from across the Atlantic Ocean caused many lapses in communication in the eighteenth century. With most of

Howe's New York troops on the Philadelphia campaign, Major General Henry Clinton, then in command in New York, tried his best to reinforce Burgoyne. Clinton's first order of business was to wait for reinforcements so he got off to a late start. Never able to make up the lost time, Clinton was unable to reach the scene of action and did not take part in the fighting.

Burgoyne began his march south in the spring of 1777, and by late June his British and German soldiers had advanced to Fort Ticonderoga on Lake Champlain. Defended by General Arthur St. Clair, the fort fell without resistance, and Burgoyne felt that his campaign had gotten off to a good start. But then he made a tactical error, the first of many. Once in command, he dispatched Brigadier General Simon Fraser to pursue the routed rebels. Similar to Lexington and Bunker Hill, the rebels inflicted serious losses on the British, losses they could ill afford as their troops could not be reinforced. In addition, many of Burgoyne's troops were committed elsewhere. Four hundred were assigned to guard the ammunition supplies at nearby Crown Point, an American fortification that he had captured on his trip south, and another nine hundred to maintain control of Fort Ticonderoga; the fixed manpower at Burgoyne's disposal was starting to diminish.

Figure 52. *General John Burgoyne*, Sir Joshua Reynolds, ca. 1766, oil on canvas, 127 × 101 cm. © The Frick Collection

The route from Fort Ticonderoga to Albany was mined with obstacles. The maps that had been consulted so trustingly in London did not show that the terrain was a nearly impenetrable wilderness that would break down the troops and give the rebels plenty of time to strengthen their forces. The British had relied on their partnership with the Indians for their offensive, but in July 1777 a renegade war party had murdered and scalped one Jane McCrea, who happened to be the fiancée of one of Burgoyne's officers. This and other barbarous actions carried out by the Indians dismayed the general, largely because they conflicted with the so-called laws of civilized warfare. Always a stickler for regulations, Burgoyne did not have success in bringing the Indians in compliance with the protocols of the royal army, and the repercussions of this murder caused many Indians to desert, weakening Burgoyne's ranks still more.

Throughout the summer of 1777, Burgoyne had been anticipating reinforcements in the person of William Howe, but on August 3 he received the news that the general was on his way to Philadelphia. That August, Burgoyne suffered another setback at the Battle of Bennington, Vermont. Led by a German officer, Lieutenant Colonel Friedrich Baum, who was well known for being unable to "utter one word of English," the encounter with John Stark of the New Hampshire militia, a capable veteran of the Seven Years' War and the Battle of Bunker Hill, resulted in a loss of almost a thousand of Burgoyne's troops, a sizable percentage of the soldiers under the general's command. At a time when the American militia numbered more than nine thousand and was gaining in troops, Burgoyne was down to seven thousand.

In September 1777, Thaddeus Kosciuszko, the Polish military engineer, played a role in taking advantage of the steadily fading Burgoyne. Under the direction of the American commanding officer General Horatio Gates, Kosciuszko selected the site known as Bemis Heights for its defensive potential and spent a week directing the army in the construction of the defensive works that he designed. Bemis Heights had a commanding view of the country and the only road to Albany, where it passed through a defile between the Heights and the Hudson. Immediately to the west lay heavily forested bluffs that would have been virtually impassable for Burgoyne's heavily equipped army.

Gates noted that "the great tactician of the campaign were the hills and forests, which a young Polish Engineer was skillful enough to select for my encampment." In the words that George Bernard Shaw placed in the mouth of General Burgoyne in his 1897 play *The Devil's Disciple*: "In five days I will be at Saratoga with five thousand men to face eighteen thousand rebels in an impregnable position."

The manuscript map illustrated here of the works at Bemis Heights (figure 53) is the only known specimen of the fortifications in the hand of Kosciuszko himself. It was recently unearthed and is presently in a private collection. The map delineates all of the major elements of the battlefield. Along the lower part of the map is Bemis Heights, occupied by the Americans on September 12, as well as the adjoining positions, taken up on September 20. To the north are the British positions, also as of the twentieth. The main road from Saratoga to Albany runs through the center of the map, with the Americans clearly blocking the British passage to their objective at Albany.

With supplies running woefully low, Burgoyne made his move on September 19. The first battle of Saratoga began with Brigadier General Simon Fraser leading a flank attack in a heavily wooded farmstead belonging to John Freeman. There, his troops met up with Continental regiments led by Daniel Morgan and Henry Dearborn, whose sharpshooters were able to kill or wound many British officers. As the battle intensified, Burgoyne dramatically entered the fray magnificently dressed for the occasion in his full general's uniform. The colorful outfit provided an excellent target, but luckily for him he was not killed. His aid, however, took a bullet and died on the battlefield.

The brilliant Benedict Arnold participated in the battle on the American side and after three or four hours the battle ended with heavy casualties on both sides. Burgoyne had lost 15 percent of his irreplaceable troops, but the battle was basically a stalemate. Burgoyne declared it a victory because he had held his ground, but his army was in no condition to continue the fight. It was Arnold who delivered the fatal blow to Burgoyne's offensive a few weeks later.

After the first battle of Saratoga, American general Horatio Gates withdrew to the Bemis Heights location that Kosciuszko had so strongly fortified. At this juncture, Burgoyne had to choose between two options: he could either withdraw from the field or make another advance on his resilient enemy. Though licking his wounds and down to only five thousand men on reduced rations, a general in the most powerful army in the world could only fight on. Eighteen days later, on October 7, Burgoyne attacked Gates's superior force of twelve thousand militia. Clinton was still three weeks away from reaching Burgoyne at Saratoga.

The initial American counterattacks quickly took advantage of Burgoyne's exhausted, dispirited, and undernourished manpower. British troops, trained to fight in open fields, were no match for the Americans in the wooded and rough terrain. The American commanders Dan Morgan and Enoch Poor forced the British to retreat. Then Benedict Arnold galloped onto the field to deliver the fatal blow of the greatest American victory of the war. Arnold had sustained a wound in his leg, but General Fraser and several other British and German officers had been killed and Burgoyne lost nearly 20 percent of his remaining troops.

Figure 53. *Plan of the Battles of Saratoga*, Thaddeus Kosciuszko, ca. 1777, manuscript, 31 × 20 cm. Private collection

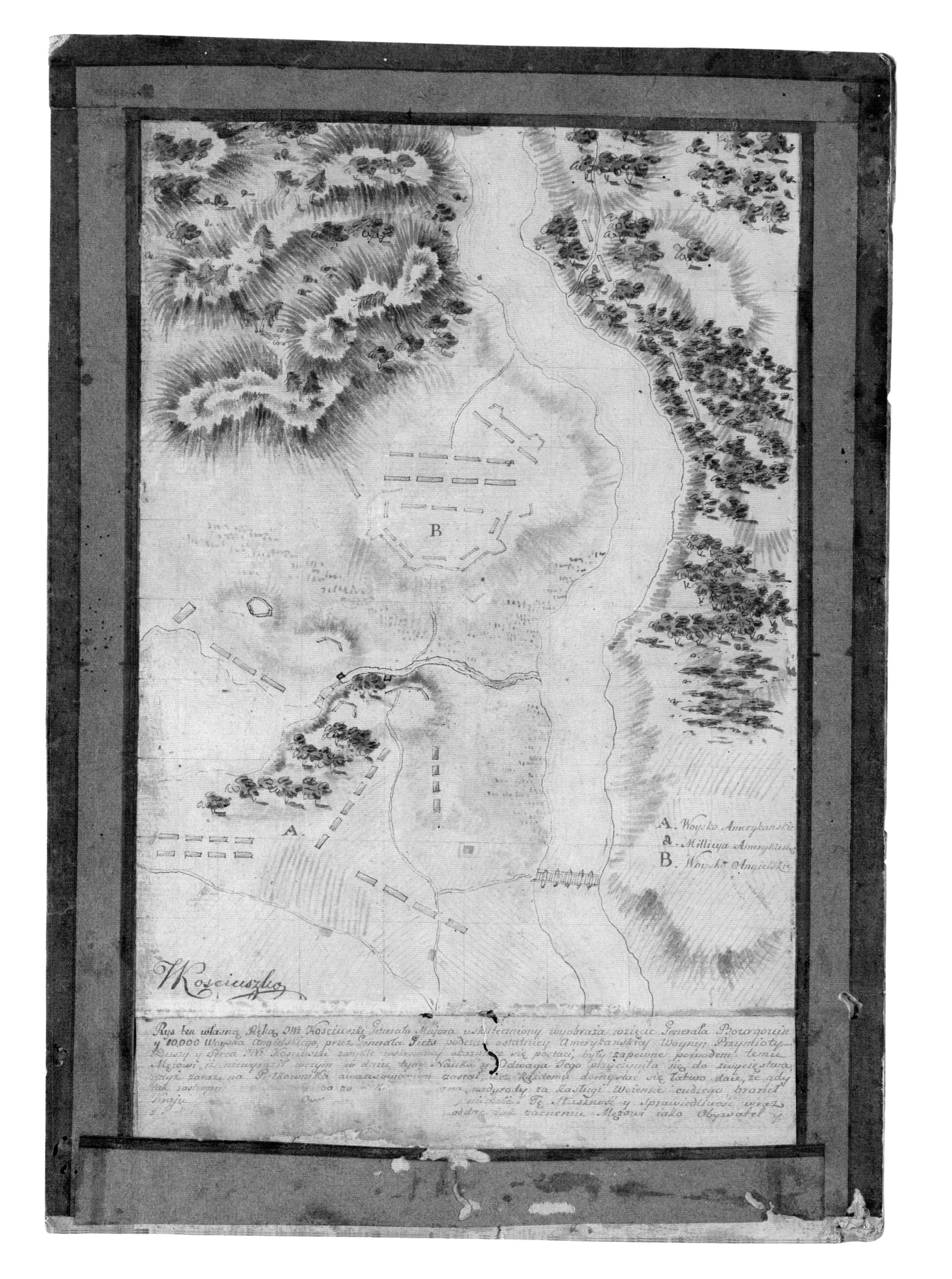

B
A.
A. Woysko Amerykanskie
a. Millicya Amerykans
B. Woysko Angielskie
TKosciuszko
Rys ten własną Ręką JWo Kosciuszki Generała Majora uskutecznіony wyobraża wzięcie Generała Bourgoin
y 10,000 Woyska Angielskiego, przez Generała Gets podczas ostatniey Amerykanskiey Woyny; Przymioty

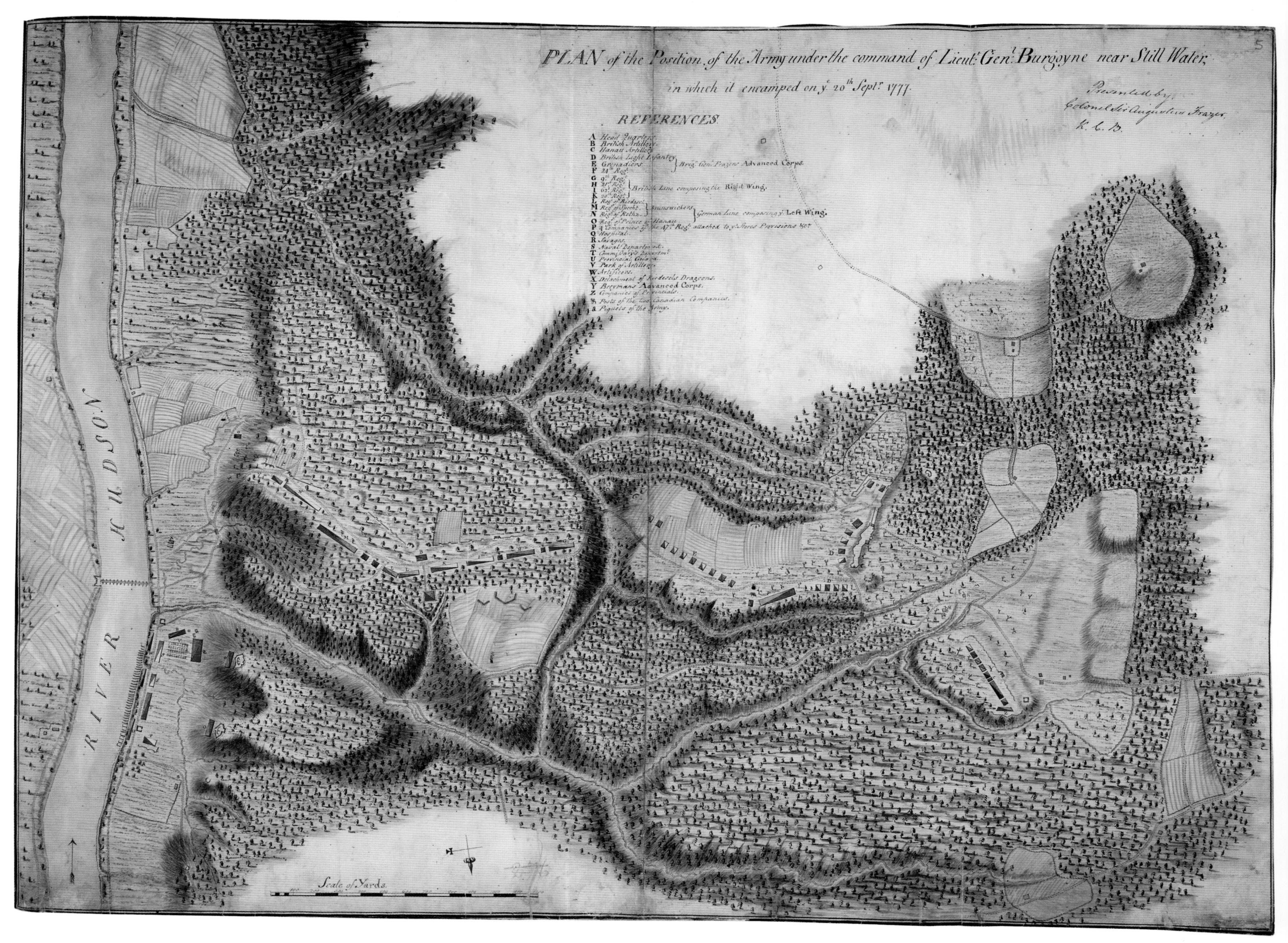

Figure 54. *Plan of the Position of the Army under the command of Lieut. Genl. Burgoyne near Still Water, in which it encamped on ye. 20th Septr. 1777*, cartographer unknown, 1777, manuscript, 53 × 71 cm. © The British Library Board [Add. 57715 (Part 5)]

The manuscript plan of Burgoyne's army near Stillwater (figure 54) along the Hudson River belonged to General Simon Fraser and was presented to the British Library by Colonel Augustus Frazer, a descendant. The map shows the bridge of rafts across the Hudson River that had been built by Fraser's advance guard. The location of this guard unit is clearly identified on the map. On October 17, Burgoyne surrendered to the Americans. More than just a victory, the battle brought the Americans aid from the French, who had been waiting on the sidelines before entering the war against their longtime enemy.

Newport

The British occupation of New York City deprived the rebels of the best port along the North Atlantic coast of America. As if that were not enough Lord Percy, under orders from General Henry Clinton, had captured Newport, Rhode Island, in December 1776 with six thousand troops. Percy remained on the scene as the commanding general for a few months, and Newport served as the second naval base for the British. Regular commerce took place between Newport and New York harbors. The British presence in these two ports was vexing to the Americans and restricted their movements along northeastern coastal waters.

In the years leading up to the Revolution, Newport had been the major city in the colony of Rhode Island. Shipbuilding and trade in rum, molasses, and slaves had made it prosperous. But once the British took charge, many of its citizens left and economic growth ground to a standstill. The last days of Newport's glory are captured in detail on Claude Joseph Sauthier's March 1777 *Plan of the Town of New Port with its Environs* (figure 55), *Survey'd by Order of His Excellency the Right Honourable Earl Percy Lieutenant General Commanding His Majesty's Forces on Rhode Island*. Largely a real estate map of recently occupied territory, it included lots within the town with the houses colored red. Docks fringe the harbor while on the outskirts streets give way to cultivated farms, which are colored green; roads are also mapped. A dispute with General Howe, the commander of all British forces in North America, over the disposition of troops led to Percy's recall to England, and on May 5, 1777, he boarded a ship in Narragansett Bay and sailed home. His mother had died during the Rhode Island campaign, leaving Percy a vast inheritance. Traveling with Percy that day was Sauthier, the talented mapmaker and his personal secretary. Newport lost its dominant foothold in Rhode Island after the Revolution as Providence gained the upper hand as the economic center of Rhode Island. Percy never returned to America.

Late in 1776 Benjamin Franklin had gone to Paris to discuss a political alliance with France. After its defeat in the French and Indian War, the country had lost not only a continent but, in addition, the capability to trade with North America. On February 6, 1778, France became the first European nation to recognize the embryonic United States, and on February 8 King Louis XVI had galvanized the fragile alliance by ordering naval commander Charles Hector, Comte d'Estaing, to sail from Toulon, France, across the Atlantic Ocean and aid the American cause. It was a well-timed moment in history. Weary of waging multiple battles at various far-flung locations on earth, Britain was showing signs of weakening. There were the possessions in India to protect, islands in the West Indies, Gibraltar, and Minorca as well as the defense of their own debt-ridden country, which the opportunistic French were belligerently eyeing.

General Washington knew that on April 13, 1778, d'Estaing had left France with a fleet of twelve warships, four frigates, a crew of 9,600, plus a thousand soldiers. The French were hoping to engage the British while they were still in Philadelphia, and the Delaware

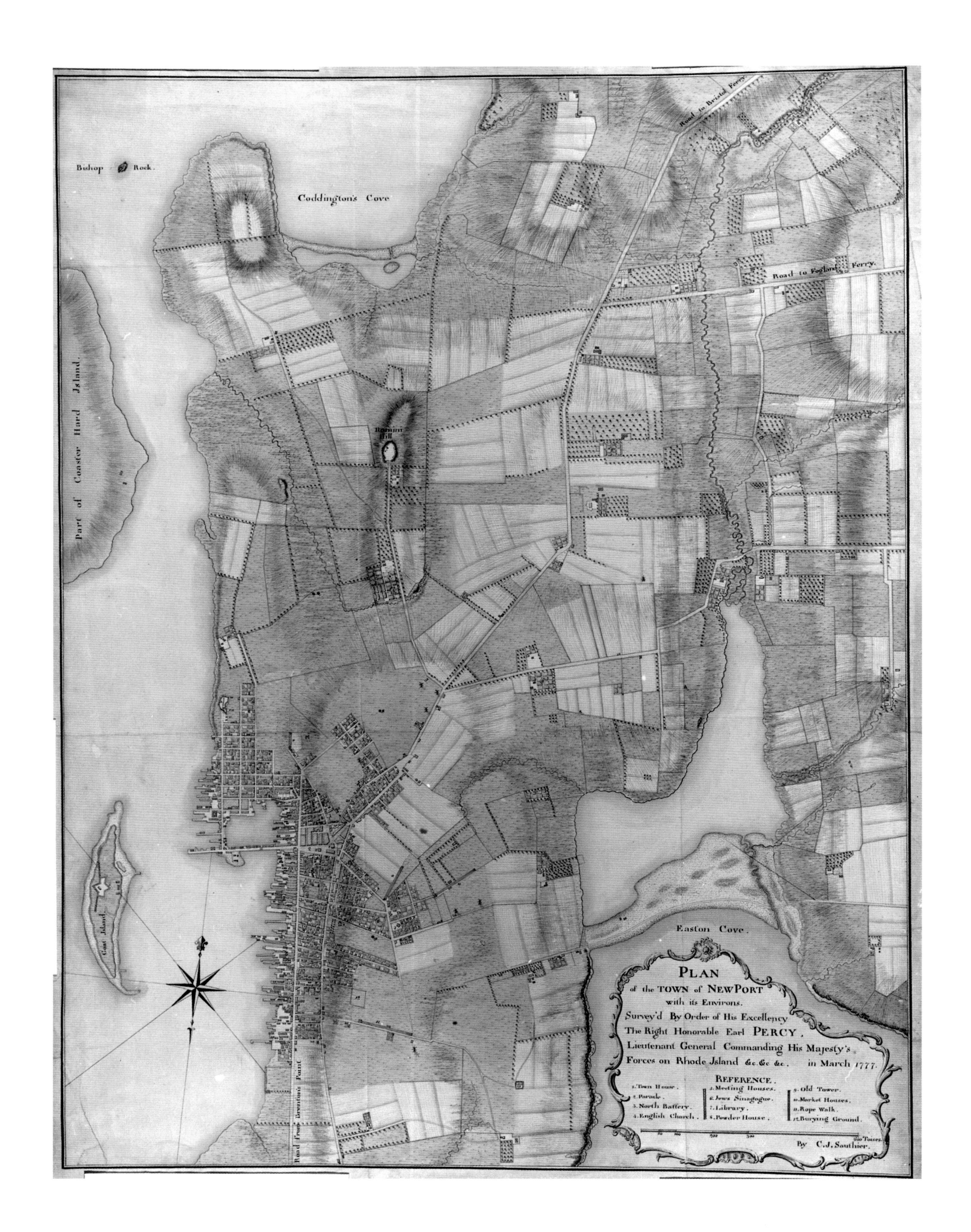
Bishop Rock.
Coddington's Cove
Part of Coaster Hard Island
Road to Bristol Ferry
Road to Fogland Ferry
Easton Cove.
PLAN
of the TOWN of NEWPORT
with its Environs.
Survey'd By Order of His Excellency
The Right Honorable Earl PERCY,
Lieutenant General Commanding His Majesty's
Forces on Rhode Island &c. &c &c. in March 1777.
REFERENCE.
1. Town House.
2. Parade.
3. North Battery.
4. English Church.
5. Meeting Houses.
6. Jews Sinagogue.
7. Library.
8. Powder House.
9. Old Tower.
10. Market Houses.
11. Rope Walk.
12. Burying Ground.
600 Toises.
By C.J. Sauthier.
Goat Island
Road from Brenton's Point

Figure 56. *Le Cte. D'Estaing laisse deux Vaisseaux et trois Frégates de son Escarde à la poursuite de la frigate Anglaise la Mairmaide*, Pierre Ozanne, 1778, manuscript, 24 × 40 cm. Private collection

Capes were d'Estaing's first destination. Leaking ships, broken sails, and stagnant wind delayed the voyage and American landfall was not sighted until the eighty-seventh day at sea. While d'Estaing was en route, the overextended British decided to evacuate Philadelphia. The city had not been a very good prize after all. On June 18 General Howe had marched his troops back to New York, so when d'Estaing reached the Delaware River on July 8 his opportunity to trap the British had vaporized. Nevertheless, a skirmish took place on July 8 at the entrance to the Delaware River. D'Estaing's fleet came across a lone, virtually defenseless English ship, the *Mermaid*, fired upon the vessel, and chased it ashore. Pierre Ozanne, the artist who accompanied d'Estaing, executed a wash drawing of the event, which is illustrated above: *Le Cte. D'Estaing . . . à la poursuite de la frigate Anglaise la Mairmaide* (1778) (figure 56). Such was the first naval engagement between the French and the English in the Revolutionary War.

Without a navy to engage on the Chesapeake, d'Estaing next headed north in quest of Admiral Richard Howe's fleet in New York's harbor. The French squadron arrived at Sandy Hook at the entrance of the harbor on July 10. The battleships were veritable behemoths loaded with as many as one hundred guns. They would have been formidable adversaries

Figure 55. *Plan of the Town of New Port with its Environs*, Claude Joseph Sauthier, 1777, manuscript, 97 × 73 cm. Collection of the Duke of Northumberland, Alnwick Castle

for Howe's smaller force but for the fact that they were too heavy to cross the bar at the entrance of the bay. The Americans could not even employ an experienced pilot to shepherd the ships through the narrow and shallow waters. To resolve the hopeless situation, d'Estaing consulted with General Washington, and they decided to abandon New York in favor of Rhode Island. The garrison at Newport became the next offensive for a joint attack. The French armada continued north. It was in Rhode Island, after two false starts, that the military action with France would finally take place.

D'Estaing is not remembered for his dexterity at sea. After lingering for eleven days in the vicinity of New York, it took another six to sail to Point Judith in Rhode Island where the resident British forces instantly went on high alert. Suffering from scurvy and a host of other nautical ailments, the health of the French sailors had declined after four months at sea.

The British garrison at Newport was under the command of Sir Robert Pigot, and he and his outnumbered three thousand soldiers at first appeared to be easy prey for the French naval and the American land forces that were assembling around the harbor. Admiral Howe set out from New York to support Pigot who was not protected by sea. A warship from Halifax joined the defensive along with another from Admiral John Byron's fleet. General Washington had ordered John Sullivan to Newport and then to occupy Providence, Rhode Island, and observe the British. The New Hampshire major general was able to recruit five thousand militia in short order from Rhode Island, Massachusetts, and Connecticut. Following on Sullivan's heels were Nathanael Greene and Lafayette with two additional brigades, increasing American forces to more than ten thousand. The plan was to trap the British at Newport.

The rebels were ready to confront the British by the first week in August. Pigot learned of the planned assault and began calling his troops to the city, abandoning such strategic nearby locations as Butts Hill. On August 8, d'Estaing boldly entered the Newport harbor and the next day began landing his four thousand soldiers on Conanicut Island, directly across the water from Newport, while Sullivan moved his army onto high ground above the city. As the combined forces took their positions to attack, Admiral Howe arrived at Point Judith in command of eight ships of the line. D'Estaing ordered his men back on board and on August 10 set off with his superior navy in pursuit of Howe.

To prevent the French from landing troops in Newport's harbor, the British began sinking transport vessels in the harbor to block access. One of these doomed ships was the famous *Endeavour*, the ship that Captain James Cook had used on his voyage to the South Pacific. In all, the British sacrificed thirteen ships in this defensive manner.

Sullivan brought his troops to within a mile of the British fortifications. On August 11, d'Estaing's ships were at full sail in pursuit of Howe, who had arranged his warships in a checkerboard formation in a demonstration of invincibility. Both d'Estaing and Howe had observed unsettled weather as they jockeyed for advantageous positions at sea, but neither could have predicted that they were sailing into one of the greatest storms of the century. The hurricane became the biggest factor in the ensuing conflict.

By late afternoon the high winds had blown the French and English ships out of each other's range, and by evening the gale-force winds were playing havoc with the navies. D'Estaing's flagship *Languedoc* suffered as much as any ship that managed to stay afloat.

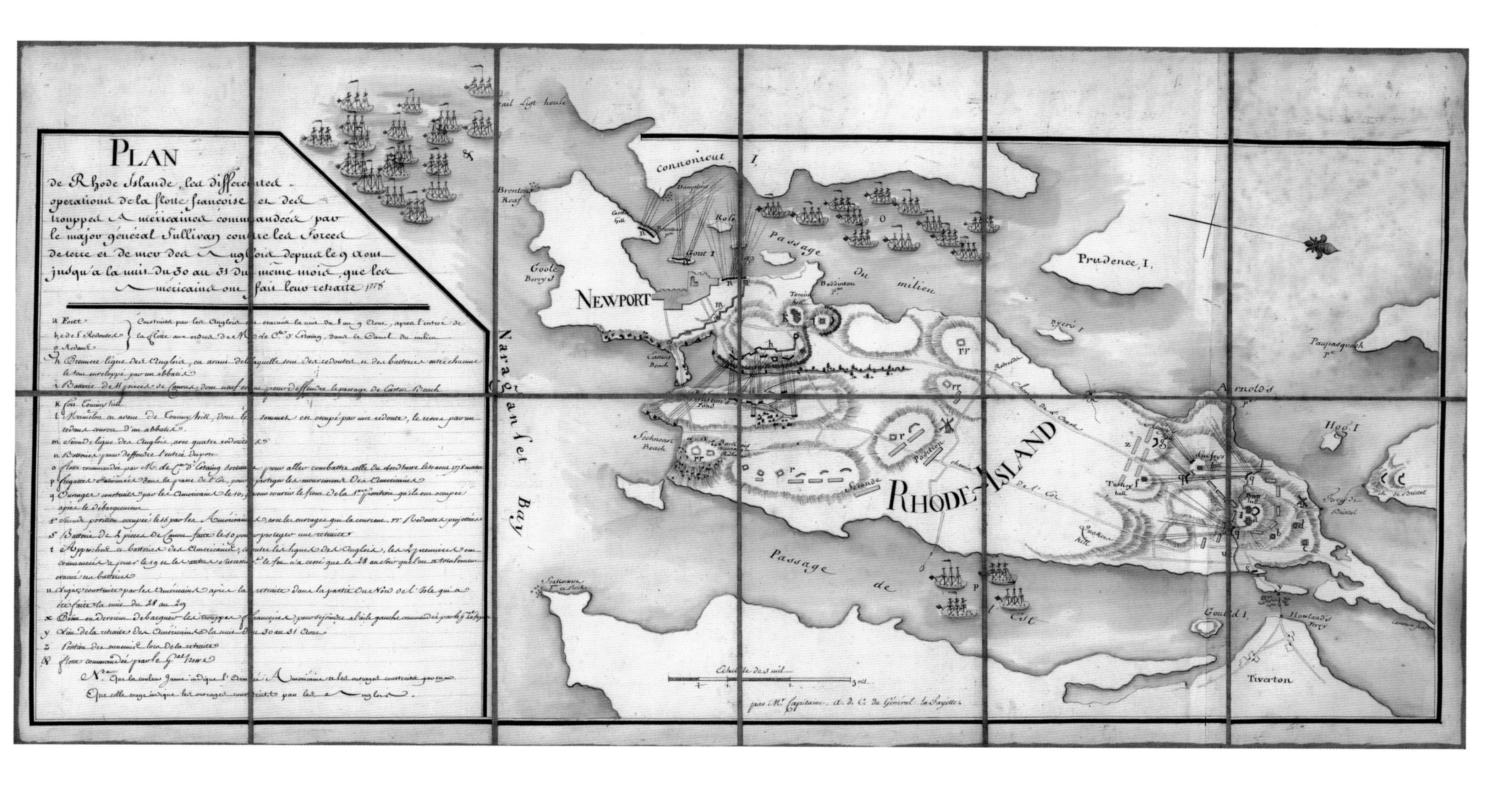

"In less than a quarter of an hour," wrote one of the sailors in his diary, "we no longer had any mast and we were as flat as a floating dock." When the rudder snapped and the anchor, with its cable attached, fell into the sea, "the skeleton of this beautiful vessel was drifting in silence at the mercy of the storm and the waves."

The British ship *Renown* had fared better than the *Languedoc* in the tempest. On August 13 the skies above Newport cleared and the *Renown* came upon the crippled *Languedoc* and fired on it. The two beleaguered vessels somehow managed to stay adrift and several French warships joined the battle, drove off the *Renown*, and rescued d'Estaing's ship. At this point d'Estaing had had enough of the French–American alliance and withdrew, much to the consternation of the American troops ready to do battle with the British. He headed to Boston Harbor on August 21 in the wake of urgent pleas to stay from various Americans and his fellow countryman Lafayette. He expected to have his ships refitted there.

Unfortunately, many of the Americans took their cue from d'Estaing and left the scene of battle. Sullivan's troop size diminished by half at the very moment when he was about to wage war on Pigot's army. The land battle that followed is well delineated on Capitaine du Chesnoy's manuscript map *Plan de Rhode Islande* (figure 57), with the English batteries in and around the Newport harbor and the location of Sullivan's army in front of Butts Hill. The Battle of Rhode Island took place on August 28 and 29, 1778, and ended in a draw with losses about the same on both sides. Capitaine's other map of Rhode Island (figure 58) shows the position of his troops and General Sullivan's as they made their retreat from the field of battle, a day before four thousand British reinforcements under Clinton

Figure 57. *Plan de Rhode Islande, les differentes operations de la flotte françoise et des troupes Américaines*, Michel Capitaine du Chesnoy, 1778, manuscript, 38 × 63 cm. Geography and Map Division, Library of Congress

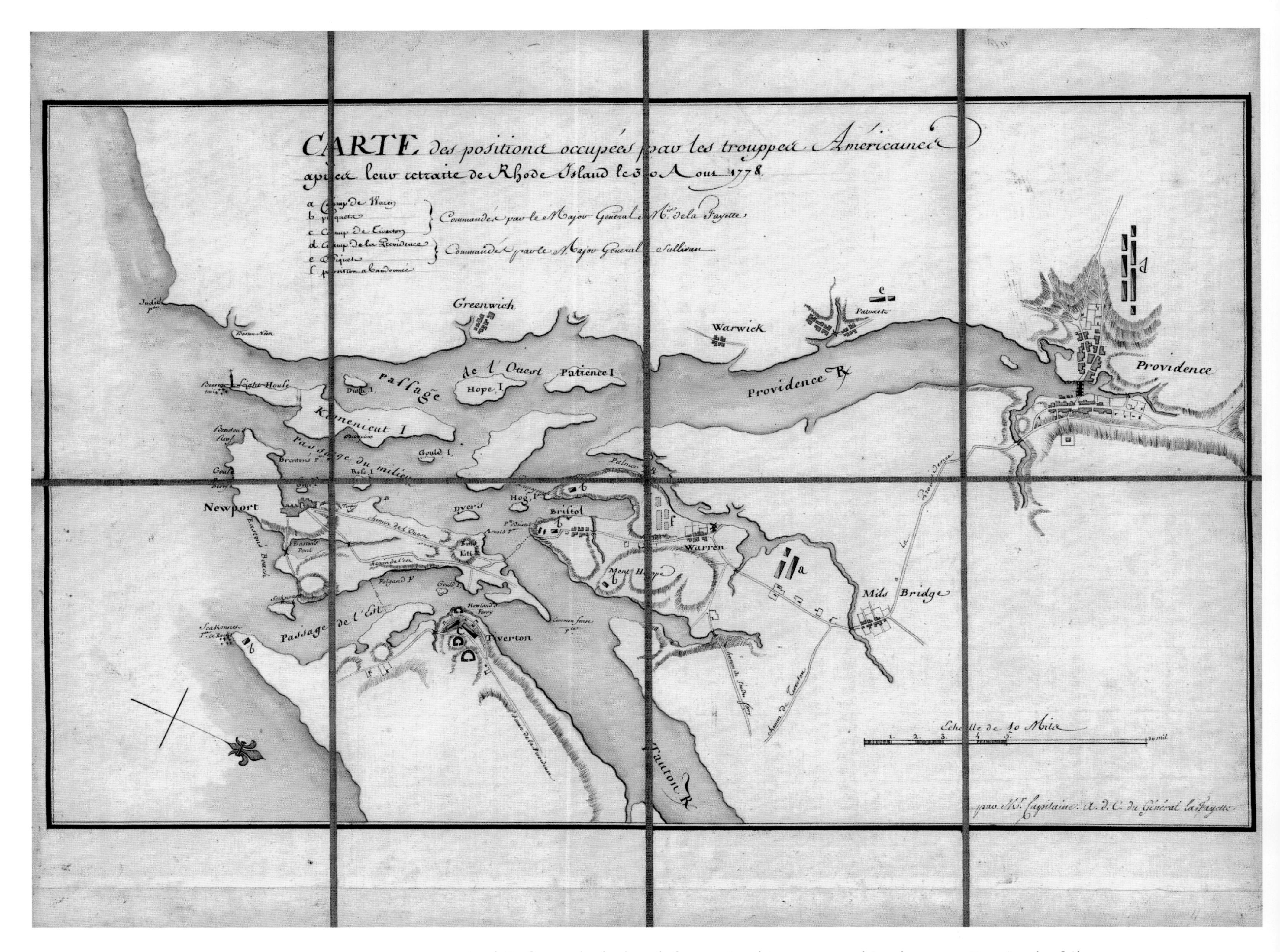

Figure 58. *Carte des positions occupeés par les troupes Americaines apres leur retraite de Rhode Island le 30 Aout 1778*, Michel Capitaine du Chesnoy, 1778, manuscript, 32 × 51 cm. Geography and Map Division, Library of Congress

arrived. Lafayette had a knack for turning his retreats to his advantage. Despite the failure of the offensive, Congress delivered formal recognition for the Frenchman's services in the successful evacuation of Rhode Island.

The first joint effort with the French ended in failure, disappointment, and stalemate. General Sullivan was furious that d'Estaing had deserted the theater of war but George Washington rose to the occasion to avoid a total collapse in French–American relations. "I will use every possible means in my power to conciliate any differences, that may have arisen in consequence of the Count d'Estaing's going to Boston," he wrote to Congress, "and I have also taken opportunities to request all the general officers here to place the matter in the most favorable point of view."

The Battle of Savannah

Following the Battle of Saratoga, warfare in the northern colonies stalled. Unable to lure Washington "into a general and decisive action," General Clinton started eyeing the south for an invasion. Not only were the southern colonies inhabited by loyalists, they were also poorly defended. On March 8, 1779, a "Most Secret" directive from Lord Germain to Clinton asserted, "The conquest of these provinces is considered by the King as an object of great importance." Clinton set his sights on Georgia where Lieutenant Colonel Archibald Campbell was already in place. On December 29, 1778, Campbell had overwhelmed the American garrison at Savannah and boasted that he had been "the first British officer to rend a star and stripe from the flag of Congress." General Washington immediately appointed Benjamin Lincoln, one of his most trusted generals, to command America's southern army. But with only two thousand troops and no navy Lincoln had little expectation of retaking Savannah.

James Oglethorpe, a British general, politician, and social reformer, had founded Savannah in 1733. Laid out on twenty-four squares as shown on the manuscript map drawn by the British officer James Moncrief (figure 59), it was America's first planned city. John Richardson, aboard the British privateer *Vengeance,* described the approach to Savannah on March 15, 1779, in a letter to his employer John Porteous: "The Banks of the River on both sides until you come near the town . . . is a Swamp. Here you can find multitudes of alligators lying in the mud like old Logs . . . The Town attends upon a steep sandy Bank, which will put a man out of breath before he can reach the Top of it." The manuscript map (figure 60) by the French engineer Pierre Ozanne provides a detailed view of the complex topography of the city's environs.

The French admiral Comte d'Estaing had sailed to the Caribbean following his disastrous campaign in Rhode Island. It was a time when France and England were struggling for control of the West Indian islands. France's colony Saint-Domingue (modern Haiti) was then the richest island in the world, and its principal city, Cap François, was the most elegant and sophisticated town in the Caribbean. The richly colored map (figure 61) by the French engineer René Phelipeau delineates the "Paris of the Antilles," where d'Estaing's fleet anchored in 1779.

General Lincoln sought out d'Estaing to help retake Savannah. The prospect of redressing the admiral's humiliating performance in Rhode Island proved irresistible, in part because his intelligence indicated that little resistance could be expected from the British garrison. He gallantly offered his full support and shortly thereafter sailed from Cap Françaís. D'Estaing's sudden appearance in Savannah astonished Georgia's royal governor James Wright. "No Man could have thought or believed," Wright wrote to Lord Germain, "that a French Fleet of 25 Sail of the line with at least 9 Frigates and a number of other Vessels would have come on the coast of Georgia in the month of September [hurricane season] and landed from 4000 to 5000 troops to besiege the Town of Savannah."

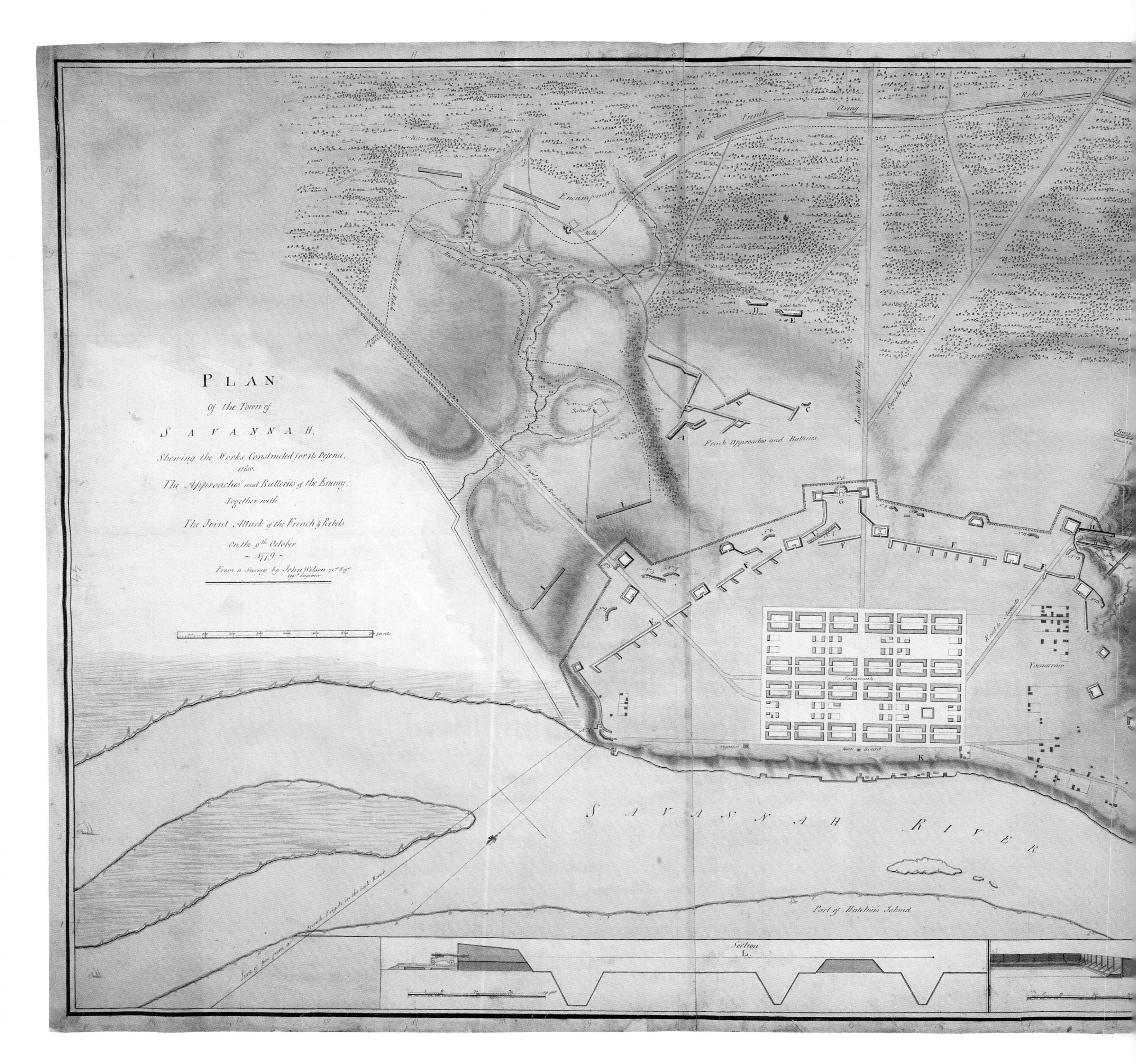
PLAN
Of the Town of
SAVANNAH,
Shewing the Works Constructed for its Defence;
also.
The Approaches and Batteries of the Enemy:
Together with
The Joint Attack of the French & Rebels
On the 9th October
~ 1779 ~
From a Survey by John Wilson 71st Regt
Asst Engineer
Encampment
the French
Army
Rebel
French Approaches and Batteries
Savannah
Yamacraw
SAVANNAH RIVER
Part of Hutchins Island
Section
L.

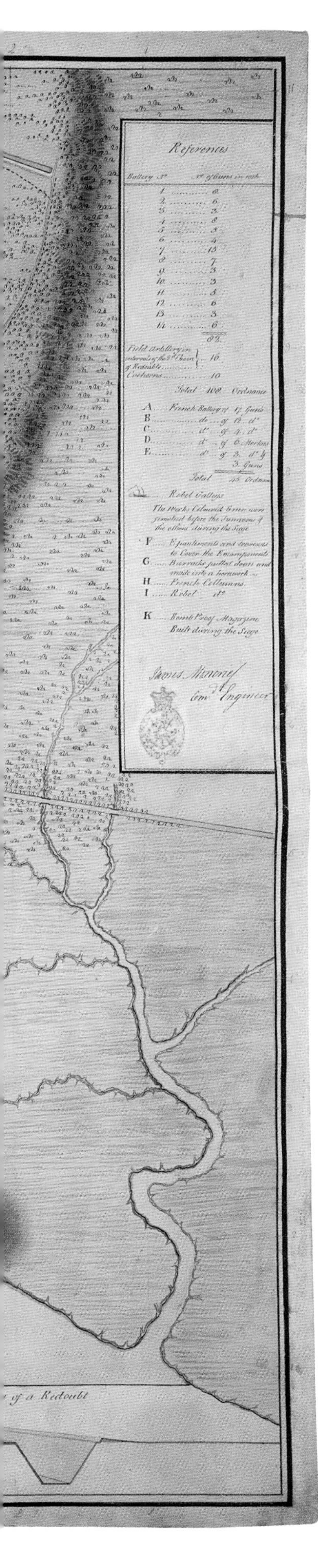

Lieutenant Colonel Augustine Prévost, Campbell's successor at Savannah, had a force of eleven hundred men but was incapable of protecting the 1,200-foot-wide land approach to the town that is shown on Moncrief's map. Fifty miles away, another eleven hundred British troops under the command of Lieutenant Colonel John Maitland were stationed in Beaufort, South Carolina. The French fleet guarded the coastal approach and Maitland's only chance to reach Savannah and reinforce Prévost was through a labyrinth of marshes and rivulets guarded by the Americans.

No Revolutionary War battle would have as diverse a cast of characters as Savannah. Traveling with d'Estaing was a long list of officers of noble birth. The force also included seven hundred free black volunteers, mostly from Saint-Domingue, and a regiment of Irish troops fighting under the banner of the French king. Lincoln's army consisted of two thousand Continentals and militia supported by the cavalry under Count Casimir Pulaski, a Polish patriot whom Benjamin Franklin had befriended and encouraged to join the American cause. On the British side, American loyalists, Cherokee braves, and two hundred armed black soldiers joined redcoats, Scottish highlanders, Hessians, and British tars. The latter removed cannon from the riverfront gunboats and manned onshore batteries.

By September 16 d'Estaing's forces were amassed outside the town and he ordered Prévost to surrender. Prévost needed Maitland's reinforcements, for without them "the only deliberation was, how to render submission as little disgraceful as possible." Prévost requested twenty-four hours to consider d'Estaing's terms. Without consulting the Americans, the admiral chivalrously granted it. It was a huge blunder, and coordination between the French and Americans, or "insurgents" as d'Estaing referred to them, remained strained throughout the campaign.

In the days leading up to September 16, the town could have been stormed and captured with ease, but on the seventeenth, d'Estaing recorded, "I have had the mortification of seeing the troops of the Beaufort garrison pass under my eyes." The resourceful Maitland had discovered an inland water passage to Savannah. British morale improved dramatically and, as Captain Charles Stedman wrote, "The presence of Maitland in whose zeal, ability and military experience so much confidence was deservedly placed . . . inspired the garrison of Savannah with new animation."

Prévost politely declined d'Estaing's terms of surrender, but the French admiral's misplaced confidence spurred him into battle. Ignoring his own deadline for the fleet's departure and believing that his superior force would quickly reduce Savannah, he started building siege trenches and brought large naval guns on shore. Over the next two weeks the French made excellent progress, but so did Captain Moncrief whose fresh redoubts appeared daily, aided by the round-the-clock work of hundreds of black slaves. The defenses consisted of a sophisticated network of redoubts

Figure 59. *Plan of the Town of Savannah, Shewing the Works Constructed for its Defense, also, The Approaches and Batteries of the Enemy, together with, The Joint Attack of the French & Rebels on the 9th October 1779*, James Moncrief, manuscript, 73 × 96 cm. © The British Library Board (RUSI MS 57716)

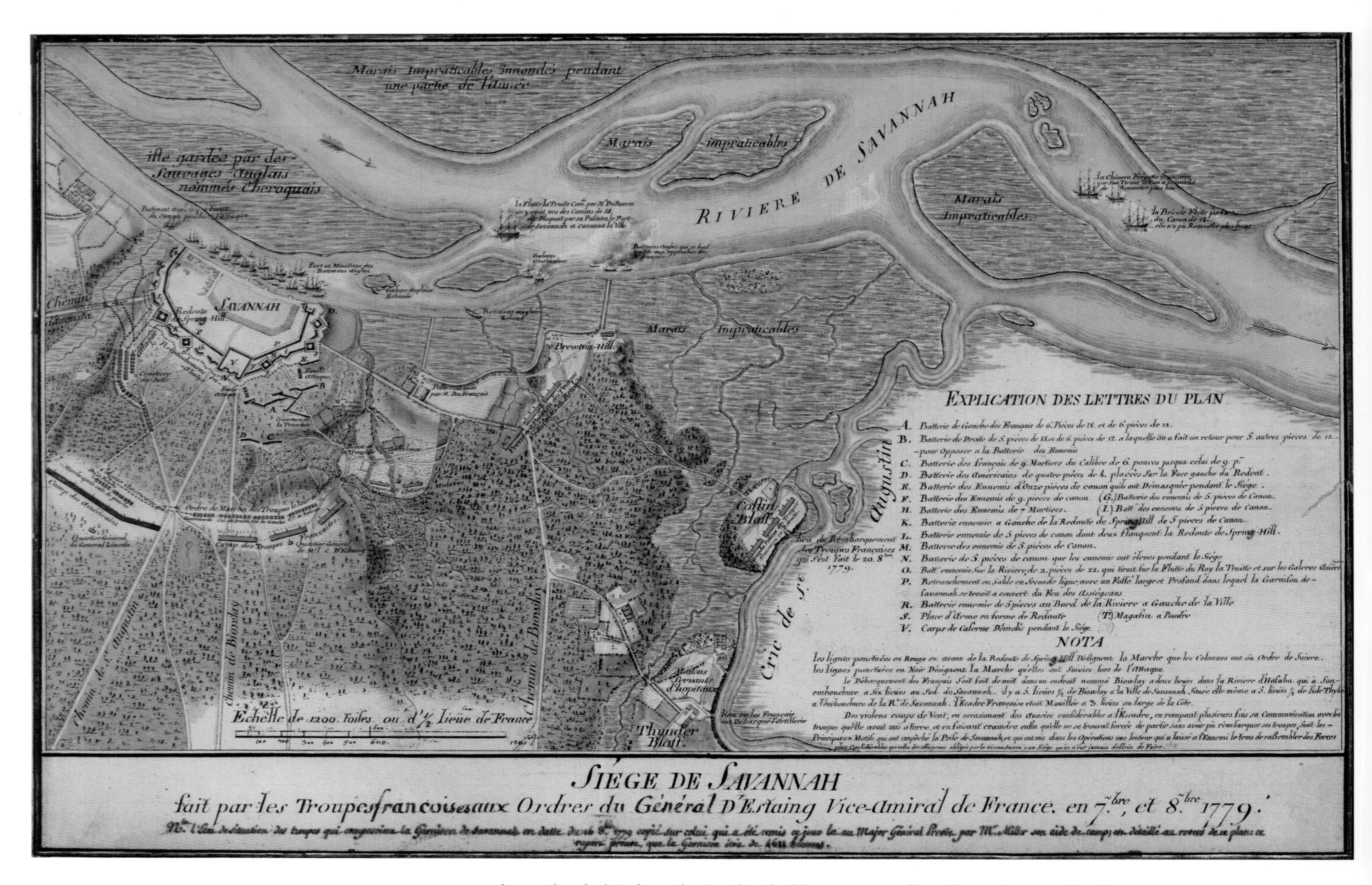

Figure 60. *Siège de Savannah fait par les Troupes Françoises aux Orders du Général d'Estaing Vice-Amiral de France*, Pierre Ozanne, 1779, manuscript, 27 × 41 cm. Geography and Map Division, Library of Congress

and trenches behind an abatis of spiked logs surrounding the perimeter. On his map, Moncrief notes that "the few Works Coloured Green were finished before the Summons & the others during the Siege."

The cannonading of the town took place from October 4 to October 8, and although considerable damage was inflicted on civilians and their property, the British troops remained protected by Moncrief's sand and palmetto log fortifications, a cross section of which appear at the bottom of the map. A more prudent officer might have withdrawn and cut his losses, but the frustrated d'Estaing feared that not attacking "would have made me a laughing stock." His decision to continue would lead to the bloodiest battle of the American Revolution after Bunker Hill.

The attack focused on the Spring Hill redoubt shown at the upper left of Ozanne's map. The French were accorded the "position of honor" and d'Estaing personally led the assault's vanguard. The attack relied on the element of surprise but, as the Frenchmen approached their objective, the predawn eeriness of Scottish bagpipes belied the fact that their plan had been discovered. As d'Estaing's grenadiers charged, they met with a galling fire, penetrating Moncrief's elaborate defensive works only to be beaten back by fire.

Figure 61. *Plan de la Ville du Cap François et de ses Environs dans L'Isle St. Domingue*, René Phelipeau, Paris, 1786, copperplate engraving, 40 × 57 cm. Private collection

Casualties from the initial charge were horrendous, made even starker by the blood-soaked white French uniforms. D'Estaing was wounded in the arm but regrouped his grenadiers for a second charge that met with no more success.

Other units including Americans under Colonel John Laurens and Pulaski's cavalry tried to take the redoubt, but they were also repelled. Adding to the confusion, the attackers ran headlong into retreating forces like a "crowd coming out of church," according to one French officer. D'Estaing suffered a second and more severe wound and had to be carried from the field. Count Pulaski was mortally wounded while leading a futile charge. The American colonel William Moultrie wrote, "They were so crowded in the ditch, and upon the berm, that they could scarcely raise an arm; and while they were in this situation, huddled up together, did the British load and fire upon them very deliberately, without any danger to themselves." As the allies commenced a disorderly retreat, Prévost ordered a sortie from the fort. Only the efforts of the French reserves, including the black regiments, slowed the British advance at the Jewish burial grounds and permitted the withdrawal of the routed grenadiers. France's engagement with the English in America had been a disaster and, as for d'Estaing, his bravery was admired but his judgment as a commander received no accolades.

Hero's tributes were bestowed on Captain Moncrief following the battle. Prévost wrote Germain, "And now my Lord give me leave to mention the great ability and Exertions of Captain Moncrief the Chief Engineer who was Indefatigable day & Night and whose Eminent Services contributed vastly to our defense and safety." Moncrief would be praised again the following year at the siege of Charleston, South Carolina.

Moncrief's and Ozanne's maps represent contrasting styles. Moncrief was an engineer working "in the trenches" and his map is appropriately detailed and utilitarian. Ozanne, on the other hand, was a marine artist and engineer traveling with d'Estaing on board his flagship *Languedoc*. His assignment was to chronicle what the French viewed as an important period in history and his map is decidedly expansive and artistic. Following the Battle of Savannah, d'Estaing returned to France, where in 1794 he became a victim of a revolution that was inspired many years earlier in America. He was guillotined.

Charleston 1780 and Camden

"I think this is the greatest event that has happened the whole war," exulted General Sir Henry Clinton when word of the victory at Savannah reached New York. With renewed optimism, he set his sights on South Carolina and put the finishing touches on his plan to invade Charleston. By the time the war ended roughly one third of all battles had taken place in South Carolina.

On the patriot side, "The most gloomy apprehensions respecting the Southern States took possession of the minds of the people," wrote the contemporary historian David Ramsey. American soldiers began to desert in droves while loyalists throughout the South were emboldened. The Savannah victory not only opened the door to the British army, it also led to bloody internecine warfare between patriots and loyalists along the southern frontier.

General Clinton and Admiral Mariot Arbuthnot sailed for Charleston on December 26, 1779. "The passage might have been expected to be performed in ten days; but such was the severity of the season that the fleet was very soon separated and driven out of its course by tempestuous weather," recalled Captain Stedman. One ship carrying some two hundred Hessian troops was blown so far off course that it landed on the Irish coast.

Clinton's armada finally assembled off the Atlantic coast near Charleston in mid-February 1780. His force of eighty-five hundred, augmented by troops from England and Georgia, had swelled to fourteen thousand men and ninety ships. Clinton had learned lessons from his unsuccessful attempt to capture the town in 1776. Rather than pursue a frontal attack on Fort Moultrie and the town's waterfront defenses, he disembarked his forces on the coast thirty miles south of Charleston and then slowly and methodically proceeded overland with the help of specially designed swamp boats. His objective was to invest Charleston from the landward side, where the Americans had given insufficient consideration to the defenses.

At Charleston, General Benjamin Lincoln commanded a force of five thousand Continentals and militia. His initial concern was less with the town's landside defenses than with keeping Arbuthnot out of Charleston's harbor. The shallow bar guarding the harbor could be passed by the men-of-war only after their cannon were removed. This process took almost two weeks during which the unarmed ships were vulnerable to attack.

The American commodore Abraham Whipple was inside the harbor with nine ships of the fledgling American navy. Whipple had a reputation as an aggressive privateer, and Lincoln expected him to challenge the British crossing. Instead, Whipple pulled back and anchored his ships in the Cooper River, as shown in the manuscript map at the Library of Congress (figure 62).

Arbuthnot's flagship *Reknown* came across the bar on March 20, and Clinton wrote the admiral, "Joy to you, Sir, to myself, and to us all upon your passage of the infernal bar." George Washington also understood the importance of passing the bar, writing John Laurens on April 26, "It really appears to me, that the propriety of attempting to defend the town depended on the probability of defending the bar." Washington had learned the efficacy of retreating to preserve his army but he also understood that Lincoln was under enormous pressure from the citizens of Charleston not to withdraw.

Clinton's army slowly worked its way to the landside of Charleston and commenced classic siege parallels moving cannon and troops ever closer to the town. The three parallels are clearly shown on the map. On April 10 Clinton's summons to surrender was refused by Lincoln. On April 14 the British cavalry under Banastre Tarleton and his green-clad Loyalist Dragoons overwhelmed an American force at Moncks Corner a few miles north of Charleston, effectively cutting off Lincoln's only line of escape. Lincoln held out until early May when he proposed terms of surrender that paroled Continental soldiers and allowed the militia to return home. Moultrie recorded Clinton's curt reply: "The alterations you propose are utterly inadmissible; hostilities will in consequence begin afresh at eight o'clock." He continued to describe the siege's last night: "The fire was incessant almost the whole night; cannon balls whizzing and shells hissing continually amongst us; ammunition chests and temporary magazines blowing up; great guns bursting, and wounded men groaning along the lines. It was a dreadful night! It was our last great effort, but it availed us nothing."

The citizens who had pleaded with Lincoln to hold fast pressured him to capitulate. On May 12, 1780, Lincoln surrendered the entire southern army, the largest single loss suffered by the Americans during the Revolutionary War. In June, Clinton returned to New York, leaving Cornwallis with an army of eighty-five hundred and instructions to "reduce" the Carolinas. No effective fighting force remained to oppose him.

Washington favored Nathanael Greene to replace the captured Lincoln as commander of the surrendered army but Congress overruled him, appointing Horatio Gates, whose reputation still sparkled from the Battle of Saratoga. Gates marched south with a core army of Continental soldiers, attracting units of militia along the way. By the time he neared

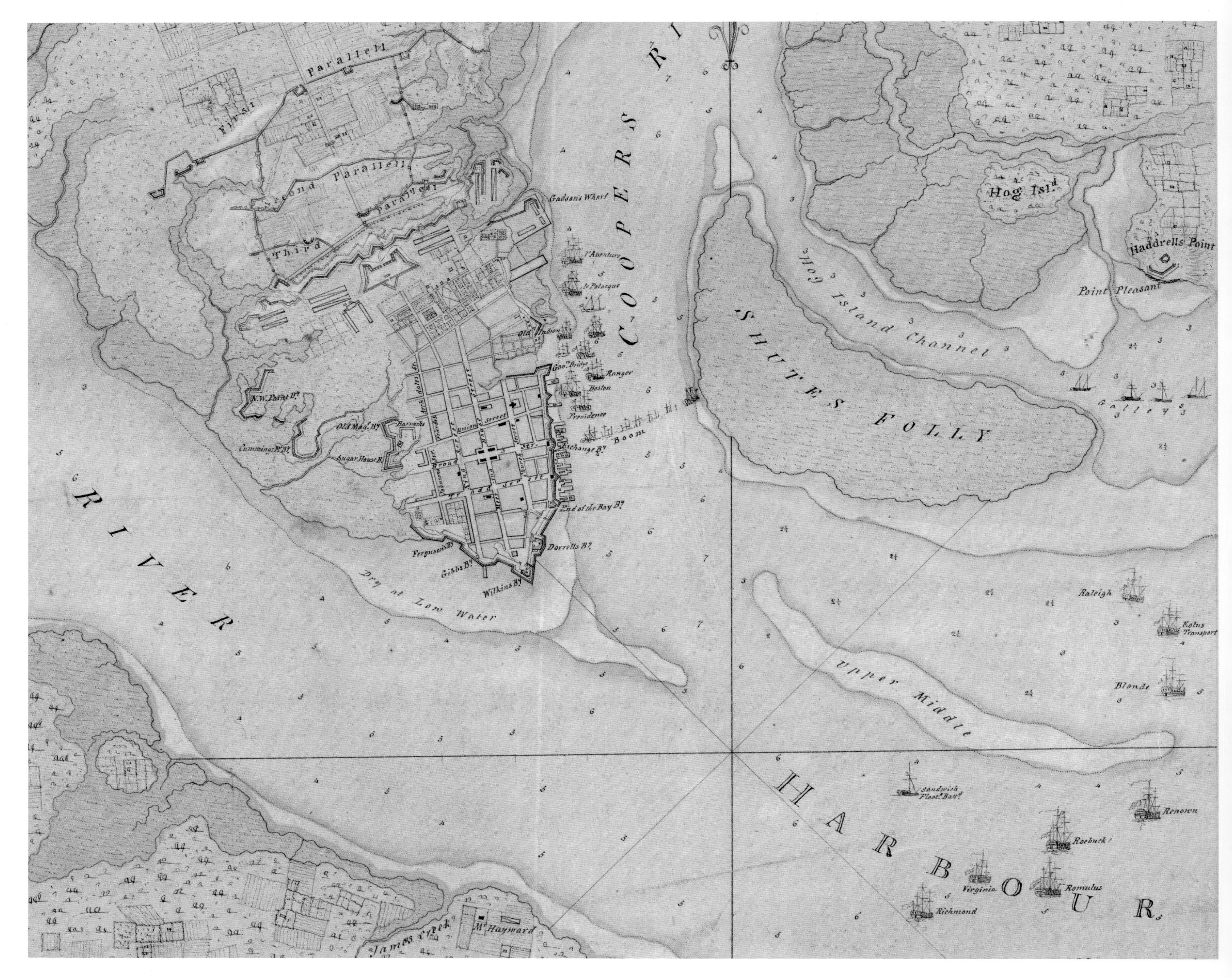

Figure 62. *The Investiture of Charleston, S.C. by the English Army, in 1780. With the position of each corps* (detail) cartographer unknown, 1780, manuscript, 70 × 51 cm. Geography and Map Division, Library of Congress

Camden, South Carolina, he commanded some three thousand troops, roughly two thirds of whom were militia. Francis Lord Rawdon with only seven hundred soldiers was holding Camden, and when Cornwallis learned of the threat he marched from Charleston to support him. The opposing forces prepared for the engagement that Edward Barron re-created on his oval map of the action (figure 63). He was not on the spot but rather drew the commemorative map for Lord Percy from official accounts provided by both Cornwallis and Gates. It is reminiscent of those illustrating European pitched battles, which is exactly what the Battle of Camden turned out to be. It is difficult to understand why Gates with

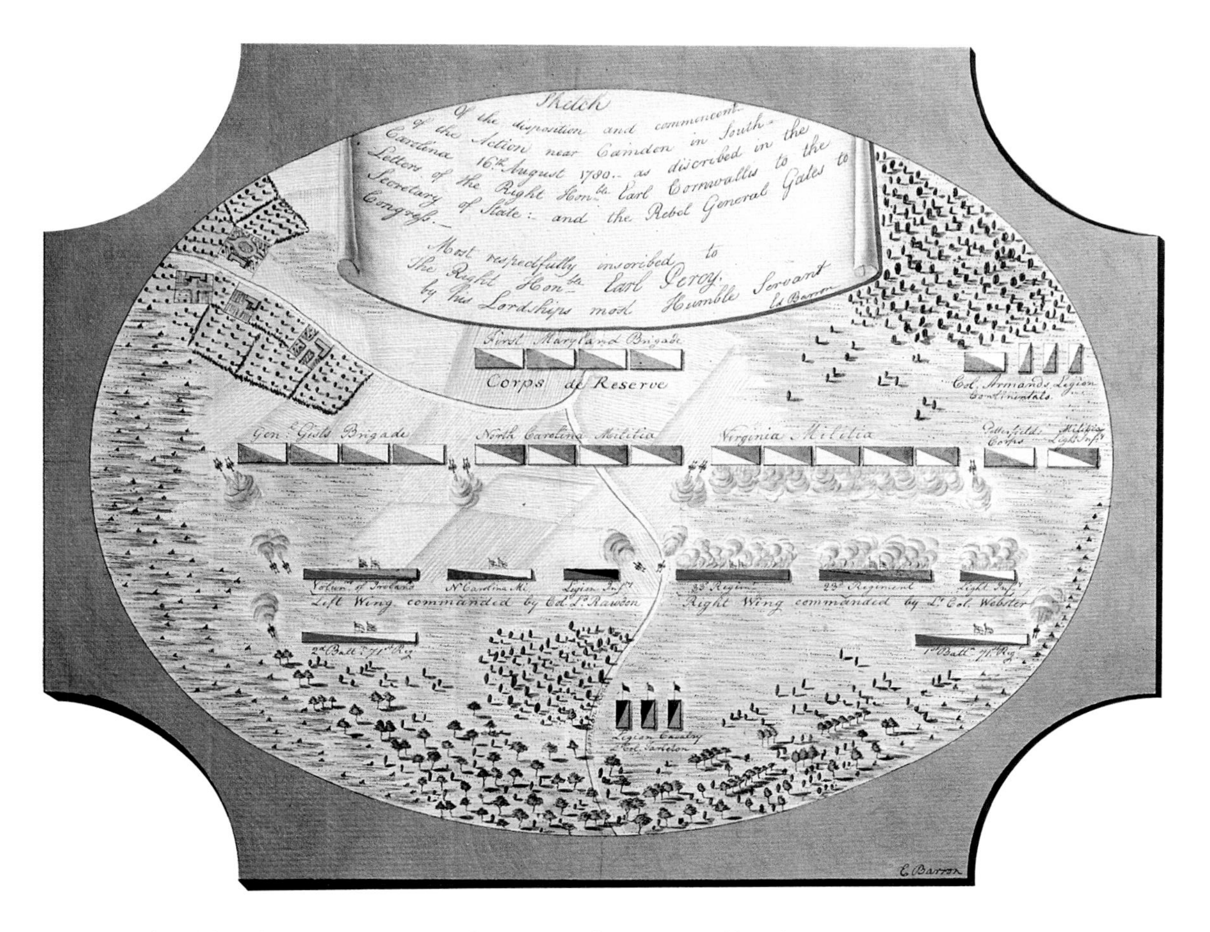

Figure 63. *Sketch of the Disposition and Commencemt. of the Action near Camden in South-Carolina 16th August 1780*, Edward Barron, ca. 1781, manuscript, 18 × 23 cm. Collection of the Duke of Northumberland, Alnwick Castle

two thirds of his force consisting of untrained militia willingly engaged in such a battle. The American militia units were placed directly across the field from the finest regiments in the British army. Gates ordered the Virginia militia to advance to open the battle but, when the British predictably charged with bayonets, they threw down their weapons and fled without firing a shot. The panic became contagious, and although the Continental forces fought bravely they were soon overwhelmed. Gates joined the retreating troops and did not stop until he was well into North Carolina.

Cornwallis had scored a resounding victory, which he and Tarleton's feared Dragoons followed up with successes throughout the Carolinas. But the British continued to lose irreplaceable troops, some from combat but many more from illness in the harsh southern climate. American privateers constantly attacked British supply lines at sea and small patriot forces harassed them along the frontiers. When Washington's man Nathanael Greene was finally appointed commander of the southern army he took a different military approach, harassing Cornwallis and then withdrawing, pulling the British farther and farther from their Charleston base. As historian Dupuy observed, "Greene's Carolina Campaign is the only known instance in military history where a General has won an extremely active and hard-fought campaign without having been able to gain a single tactical victory." Greene's actions inevitably led Cornwallis to Yorktown. The British never fully understood that, although their powerful naval and land forces could capture and hold major coastal towns, it was yet impossible to control the vast colonial interior.

Rochambeau's March

Promoted to lieutenant general for the assignment, the Comte de Rochambeau and his armada arrived in the area of Newport, Rhode Island, on July 10, 1780. Newport had been the site of the first joint operation between the French and Americans, but the effort had been a failure that had forced the Americans from that battlefield in the summer of 1778. A year later, the British, wanting to concentrate their military might, had abandoned Newport; British troops from Newport regrouped in New York City. So when Rochambeau anchored his ships in Narragansett Bay he found an undefended town. One of the first orders of business for the Royal Corps of Engineers was a survey of the defenses still in place from the previous warfare.

Newport served as the base of operations for the French fleet while it occupied the town from July 1780 until July 1781. At intervals, the French refreshed the troops as vessels from France sailed into the harbor. At the beginning of October, for example, Louis-Alexandre and Charles-Louis Berthier arrived in Newport. These brothers were the sons of J. B. Berthier, the chief French engineer during the Seven Years' War. They were mapmakers who grew up in the Dépôt de la Guerre and had studied their craft at Versailles. Louis-Alexandre, who went on to become Napoleon's chief of staff, held Rochambeau in the highest esteem and regarded him almost as a father. When the talented Berthiers arrived in America, Rochambeau realized that two of the finest French cartographers had joined his army, and right away he put them to work surveying.

While in residence in Newport during the late autumn and early winter, one of Louis-Alexandre Berthier's projects was to survey the island of Aquidneck, where Newport is located, and the surrounding area. The large map that resulted is not illustrated here, but a smaller derivative map with the title *Plan de la Ville, Port et Rade de Newport* is pictured in figure 64. It is not dated but was Rochambeau's personal copy. In 1883, the Library of Congress purchased many maps that had belonged to the commander in chief of French forces in America; it is now one of the finest of Revolutionary War map collections.

Both maps are based not only on the Berthiers' new surveys but also on maps that had been previously drawn and were on file in the Dépôt de la Guerre. Rochambeau's staff had mined the depot for relevant cartographical resources in preparation of their trip to America, and these had been put to use by the Berthiers and others. There was no military or naval action in Newport during these months of French occupation. The French soldiers and officers exercised strict discipline but it was mainly for show—while they were in Newport, they were largely idle, but their presence was important. Boston and Newport were the only two ports along the Atlantic coast that could accommodate the growing French fleet and it was necessary for Rochambeau to protect the important harbor.

And protect it he did. One look at the Berthiers' map indicates the would-be belligerence of the French soldiers. The positions of the French battleships are shown as well as artillery installations on land. The range of the guns is shown in red lines—if they were all fired at

Figure 64. *Plan de la Ville, Port et Rade de Newport, avec une Partie de Rhode-Island occupée par L'Armée Française aux ordres de Mr. le Comte de Rochambeau et de L'Escadre Française commandée par Mr. le Chr. Destouches,* Louis-Alexandre Berthier, 1781, Manuscript, 58 × 61 cm. Geography and Map Division, Library of Congress

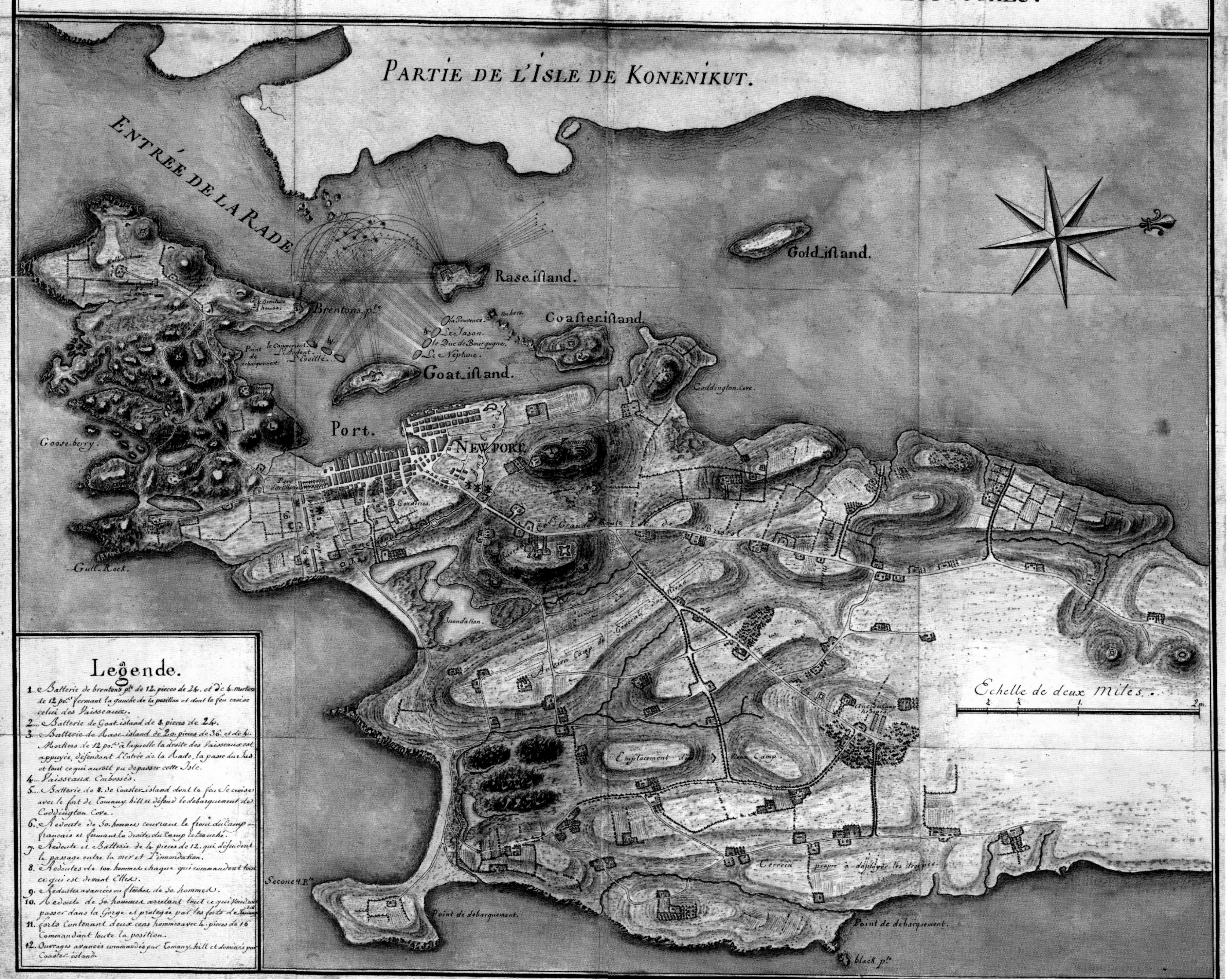
PLAN DE LA VILLE, PORT ET RADE DE NEWPORT, AVEC UNE PARTIE DE RHODE-ISLAND OCCUPÉE PAR L'ARMÉE FRANCAISE AUX ORDRES DE M.R LE COMTE DE ROCHAMBEAU ET DE L'ESCADRE FRANCAISE COMMANDÉE PAR M.R LE CH.R DESTOUCHES.
Map Division Library of Congress
PARTIE DE L'ISLE DE KONENIKUT.
ENTRÉE DE LA RADE
Gold island.
Rase island.
Brentons p.te
Coaster island.
Goat island.
Port.
NEWPORT
Coddington Cove.
Goose berry.
Gull Rock.
Inondation.
Echelle de deux Miles.
Point de debarquement.
black p.te
Legende.
1... Batterie de brentons p.te de 12. pieces de 24. et de 4. mortiers de 12 po.ces fermant la gauche de la position et dont le feu croise celui des Vaisseaux.
2... Batterie de Goat-island de 8. pieces de 24.
3... Batterie de Rase-island de 20. pieces de 36. et de 4. Mortiers de 12 po.ces à laquelle la droite des Vaisseaux est appuyée, défendant l'Entrée de la Rade, la passe du Sud et tout ce qui auroit pu depasser cette Isle.
4... Vaisseaux Embossés.
5... Batterie de 8. de Coaster-island dont le feu se croise avec le fort de Tomany-hill et défend le debarquement de Coddington Cove.
6. Redoute de 50. hommes couvrant le front du Camp françois et fermant la droite du Camp retranché.
7. Redoute et Batterie de 4. pieces de 12. qui défendent le passage entre la mer et l'innondation.
8. Redoutes de 100. hommes chaque qui commandent tout ce qui est devant Elles.
9. Redoutes avancées ou fléches de 50. hommes.
10. Redoute de 50. hommes arretant tout ce qui voudroit passer dans la Gorge et protegée par les forts de Tomany-hill
11. forts Contenant deux cens hommes avec 4. pieces de 16 Commandant toute la position.
12. Ouvrages avancés commandés par Tomany-hill et dominés par Coaster-island.

Figure 65. View of Newport, Rhode Island and the harbor showing the position of the French fleet and troop encampments (detail), artist unknown, watercolor painting, 1780, 24 × 122 cm. Private collection

Figure 66. *Camp à Baltimore*, in Rochambeau *Amérique Campagne*, 1782, manuscript, 41 × 18 cm. Geography and Map Division, Library of Congress

once, they would create the densely intersecting lines that are shown on the map. Such intense gunfire would be fatal to any ship attempting to enter the harbor, but no enemy ship tried to penetrate these trapezoids of gunfire. At this time, an unknown officer on Rochambeau's staff executed the first French view in watercolor of the town and its harbor, a detail of which is presented in figure 65. The configuration of the ships in the harbor is identical to those shown on the map of Newport (figure 64).

On May 22, 1781, Washington and Rochambeau had met in Wethersfield, Connecticut. Attacking New York City was the primary subject of this meeting, but at the beginning of June Clinton had received intelligence about the plan in the form of an intercepted document. Clinton built up his force to between fifteen thousand and seventeen thousand fighting men. By the end of July Washington realized that the New York garrison was simply too strong to attack. Then, in mid-August, Washington and Rochambeau received a letter from the Comte de Grasse informing them that he was leaving the West Indies bound for the Chesapeake. The venerable Cornwallis became the focus of joint military action, but first Washington and Rochambeau had to get their troops from New York and Rhode Island to Virginia. After many months of inactivity Rochambeau's army was ready to make its move.

The French forces began their more than six-hundred-mile trek to Virginia on June 10, 1781. Rochambeau divided the army into four divisions, each leaving a day apart. They continued across Connecticut to New York until they reached Dobbs Ferry. There, Washington and the Continental Army were encamped in July and August 1781. On August 19, the combined forces of Rochambeau and Washington began traveling south. They marched through New Jersey, Pennsylvania, Delaware, and Maryland. In 2009, President Barack Obama signed into law an act that designated the Washington-Rochambeau route a National Historic Trail.

The Berthier brothers made maps of the towns and campsites along the way. The collection that resulted not only provided a detailed narrative of the trip, they also delineated many towns and areas for the first time. The map of Baltimore that is illustrated here is not the first of the city, but it is one of the most captivating of the eighteenth century and reflects the good feelings the French had about it (figure 66). "Our camp was erected on a charming site in the midst of a woodland near the city, from which we had a most agreeable view." The Berthiers' map of Baltimore shows a street pattern near the port that exists to this day; it is in Rochambeau's journal at the Library of Congress.

On September 21 most of the army that had marched from Newport or Dobbs Ferry boarded ships in nearby Annapolis, Maryland, to carry them the last leg of the trip to the James River in Virginia.

20.d Camp à Baltimore le 24 Juillet 13 Milles 1/2
de Spurier's Tavern . Sejour Jusquau 24 Aoust

Rade

Baltimore

Port

Ferry
Bran

Patapsco R.

Yorktown

In 1956, Professor George W. Pierson was carrying out research in Paris when he met Mme F. Roux-Devillas, a descendant of General Lafayette. She was in possession of a large map of a military campaign that had taken place in Virginia in 1781, which she showed to Pierson, the chairman of the history department at Yale University. Entitled *Campagne en Virginie du Major Général M. de la Fayette*, the map chronicled many of the events leading up to the defeat of the British at Yorktown that ended the Revolutionary War (figure 67). Pierson instantly cabled Alexander Vietor, the curator of maps at the Yale University Library, with the news of this remarkable discovery, at the same time informing his colleague that the map was for sale. Vietor replied "in great excitement and told [Pierson] to close the purchase forthwith." It was "in the nature of a national treasure," exclaimed Vietor, who bought the map for the university for $1,400.

Vietor realized that the map was more than just a relic from the war. "The importance of the new Yale acquisition lies in the fact that there are few maps of entire campaigns of the Revolutionary War period surviving today." Vietor continued, "I don't know of any other Revolutionary War campaign which was so well illustrated by a single map. Unlike the Civil War the Revolutionary period produced few maps of entire campaigns and nearly all of these are British."

On April 28, 1780, General Gilbert du Motier, the Marquis de Lafayette, made a triumphant return to America. He had already served General George Washington in various capacities on his first tour of duty, but then he and his aide-de-camp Michel Capitaine du Chesnoy had taken a leave in France. They sailed back on board one of the king of France's frigates, along with a document bearing official news that a French expeditionary force under the Comte de Rochambeau was on its way to America to serve as an auxiliary to the Continental Army. As one historian wrote, "Lafayette had become one of the central figures in an alliance that was to prove decisive in the outcome of America's long war for independence. The close relationship he had established with Washington continued as Lafayette served as interpreter and liaison during Washington's meetings with the French commanders."

Throughout the war, the British had occupied New York City and their stronghold there had caused General Washington continual aggravation. In the early months of 1781 the general's military objective was New York as he coordinated his offensive with the French. At this time, British forces in the Carolinas and Virginia were increasing in number, and the theater of war was migrating from north to south. The new focus of the British campaign became the conquest of Virginia. Benedict Arnold, on his first assignment after defecting to the British, took charge of sixteen hundred British forces as they traveled from New York to Virginia. Under orders from General Clinton, he carried out raids along the way; in Richmond, for example, he burned buildings and destroyed crops.

French, British, and American forces were converging on Virginia by land and by sea. On February 20, 1781, Washington ordered Lafayette to the colony. He arrived on April 21, just a few days before General Cornwallis, whose instructions from the Royal Navy were to secure a harbor in the Chesapeake Bay and cut off supplies to the American army. During the summer of 1781, Cornwallis made the village of York (now called Yorktown), Virginia, his base of operations. Accordingly, he disembarked his six thousand regulars along with two thousand marines and loyalists in York and Gloucester, towns on opposite sides of the York River.

Scoring the decisive victory over the Americans had fallen squarely onto Cornwallis's shoulders. Under the watchful eye of Lafayette, he began to fortify the peninsula as a naval station, taking up supporting positions in Gloucester. Lafayette and his troops were no match for the English general, but the French officer provided Washington with some fortuitous geographical observations. The British were surrounded by water on three sides! The side of the peninsula connected by land proved to be the Achilles' heel as it provided access for attack. In Lafayette's eyes, Cornwallis's fortress was more a trap than a safe haven.

Campagne en Virginie gives an account of Lafayette's strategy in Virginia. While Washington and Rochambeau were marching toward Virginia, Lafayette was successfully able to confine Cornwallis to his fort by blocking his one way out: the side of the peninsula accessible by land. A French fleet with twenty-nine warships and three thousand troops under François-Joseph Paul, Comte de Grasse, sailing from the West Indies, was also en route to the Chesapeake Bay for action against Cornwallis. By October 1781 French land troops under Rochambeau and the naval forces under de Grasse had arrived, and Washington, who had traveled south with Rochambeau, was himself on the scene.

With the Americans taking positions on the right of Yorktown and the French on the left, the allied forces began digging trenches six hundred yards from the British fortifications so as to surround the besieged city. The parallel (trench) was opened on October 6, and on October 10 a bombardment destroyed artillery within Yorktown. On that day a second parallel was started three hundred yards closer to the British lines, forcing Cornwallis to abandon his outer fortification and retreat. Washington then ordered an attack on the two British redoubts, with Lafayette directing the rear action. As Rochambeau described it to de Grasse, "The smallest of these two works has been taken by the Americans under M. Le marquis de La Fayette, and it contained the battery that was the most dangerous in the York River." To avoid defeat, Cornwallis attempted to escape with his army to the nearby town of Gloucester but a storm ruined that plan. British naval forces under General Henry Clinton had sailed from New York to support Cornwallis, but their arrival at the entrance of Chesapeake Bay was too late to be of any help. Unaided and completely surrounded, Cornwallis surrendered on October 19, 1781.

One of the best descriptive maps of the engagement at Yorktown is a manuscript at the British Library by an American, Major Sebastian Bauman (figure 68). Created at General Washington's request, and dedicated to him, it was based on surveys of the grounds made a week after the battle by a military engineer who had been born and trained in Germany. A printed version was subsequently published in Philadelphia in 1782, becoming the first of the battle published in America. In addition, it has the distinction of being the first map to portray the American flag. The printed map illustrated here is at the New-York Historical

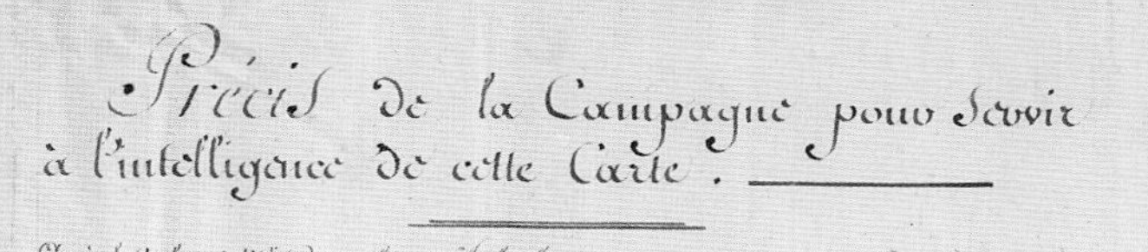

Précis de la Campagne pour servir à l'intelligence de cette Carte.

Après le Combat de Mrs. Destouches et Arbuthnot on abandonna le Projet sur Portsmouth; les Français firent voile pour Rhode Island; les milices furent congédiées, les troupes reglées se portent vers le Nord, Arnold fut ensuite renforcé par le Major Général Philipps, et la conquête de Virginie devint pour les Anglois l'objet de la Campagne.

L'Armée alliée sous les Généraux Washington et Rochambeau se portoit devant Newyork celle du Général Greene attaquoit les postes laissés en Caroline, l'une et l'autre à près de 500 milles de Richmond, le Major Général Marquis de la Fayette est chargé de deffendre la Virginie.

Avril et May.

D'après les preparatifs faits à Portsmouth il juge que la Capitale en est l'objet — marche forcée de son Corps de Baltimore à Richmond environ 200 milles, il arrive le soir du 29 Avril; les Ennemis etoient parvenus à Osborn, les petits corps de milices se rassemblent dans la nuit à Richmond, le lendemain matin les Ennemis à Manchester se voyent prevenus se rembarquent à Bermuda hundred, et redescendent James River.

Les Americains à Bottoms Bridge, un Corps detaché sur Willamsburg; le Général Philipp's recoit un aviso et remonte la Rivierre; debarquement à Brandon, second renfort de Newyork; — Lord Cornwallis qu'on assuroit etre embarqué pour Charlestown, s'avance à travers la Caroline du Nord.

Les Americains à Osborn pour etablir une Communication sur James et Appomatok Rivs. sont prevenus par la marche de Phillip's à Petersburg — le 10 à Wilton — le 18 Cannonade et reconnoissance sur Petersburg, qui en rassemblant les partis ennemis, permet de faire filer un Convoy pour la Caroline — le 20 à Richmond, Jonction de Lord Cornwallis avec les troupes à Petersburg — la grande disproportion du Corps Americain, l'impossibilité de Commander les Rivierres Navigables et la necessité de garder le coté important de James River, ne permettent pas de s'y opposer.

Ayant envoyé une partie des troupes à Portsmouth, le Lieutenant Général Lord Cornwallis se choisit une armée d'environ 5000 hommes, 300 Dragons, 300 Chasseurs montés; passe à Westover; les americains n'avoient que 3000 hommes environ, formés de 1200 hommes reguliers, dont 50 Dragons et 2000 Miliciens, tout ce que Richmond avoit d'important etait evacué, nos troupes à Wintson's Br. — marche rapide des deux Corps; les Ennemis pour engager une action les americains pour l'eviter et conserver le haut du pays, avec la Communication de Philadelphia, elle etoit egalement necessaire à notre armée et à la subsistance de celle de la Caroline.

Juin.

Les Magazins de Fredericsburg sont evacués — les americains à Mattapony Church — l'Ennemy à Chester field tavern — grandes pluyes qui sont rendre le Rapid Ann impassable — Lord Cornwallis marche pour en gagner la tête — nos troupes se hâtant et sont à Racoon ford attendre le Général Waine avec un Corps reglé de Pensylvaniens.

Desesperant d'engager une action ou de couper la Communication avec Waine et Philadelphia, Lord Cornwallis change d'objet et cherche à detourner celui des Americains, il se dirige tout à Coup sur les grands Magazins d'Albermale Court ho. — un detachement de Dragons tache d'enlever l'assemblée de l'Etat à Charlotte-Ville, et manque son Coup — un autre detachement se porte sur Point of Fork, ou le Général Stubens formoit six à sept Cens recrües, il evacüe ce point, et croit devoir se retirer dans la direction de la Caroline — quelques effets peu importants sont detruits.

Le 12 les Americains à Boswell's tavern, le passage du Rapid Ann avoit eté necessaire pour ne pas etre acculé par Lord Cornwallis: la Communication avec Philadelphia etoit indispensable; on ne pouvoit esperer même en Combattant d'empecher la destruction des Magazins avant la jonction avec les Pensylvaniens. le Général la Fayette prend donc le parti de les attendre, et dès leur arrivée regagne les Ennemis à marche forcée. Lord Cornwallis etoit parvenu à Elke Island. pour se placer au dessus des Ennemis, la routte Commune passe à la tête de Byrd's Creek, Lord Cornwallis y porte son avant garde, et Compte tomber sur notre flanc, les Americains reparent dans la nuit un Chemin peu Connu, et derobant leur marche prennent une position à Michunk Creek, ou suivant l'ordre donné ils sont joints par 600 Montagnards. le Général Anglois voyant les Magazins Couverts se retire à Richmond, et est suivy par notre Armée.

Differentes Manoeuvres des deux armées — les Americains sont rejoints par le Général Steubens avec ses recrües; leur force alors est de deux Milles hommes de troupes reglées, et 3200 Miliciens — Lord Cornwallis croit devoir evacuer Richmond — le 20 le Marquis de la Fayette le suit et Conserve l'offensive, cherchant à manoeuvrer et evitant de combattre — les Ennemis se retirent sur Willamsburg à six milles de cette Ville, leur arriere garde est attaquée avec avantage par notre Corps avancé sous le Colonel Butler — Position prise par les Americains à une marche de Willamsburg.

Juillet.

Differents mouvements qui finissent par l'evacuation de Willamsburg — les Ennemis à James Town — notre Armée s'avance sur eux — le 6 Combat vif entre l'armée Ennemie et notre avant garde sous le Général Waine en avant de Green Spring, deux pieces de Canons restent en leurs mains, mais ils sont arretés par un renfort d'Infanterie legere — la même nuit ils se retirent sur James Island — ensuite à Cobham sur l'autre Coté de James River, et de la dans leurs Ouvrages à Portsmouth.

Le Colonel Tarleton est detaché dans le County d'Amelia — les Généraux Morgan et Waine marchent pour le Couper — il abandonne son projet, brule ses Chariots, et se retire précipitamment. les Ennemis se tenant dans Portsmouth l'armée Americaine prend une position saine sur Malvan hill — se repose de ses fatigues.

Aoust.

Les Americains se refusant à descendre devant Portsmouth, une partie de l'Armée Angloise s'embarque et se rend par eau à York Town et à Gloucester. le Général la Fayette prend une position à la fourche de Pamunkey et Mattapony River; ayant des Corps detachés sur les deux Cotés d'York River, les Pensylvaniens et quelques nouvelles levées ont ordre de rester sur James River, et se croyent destinés pour la Caroline — rassemblement des milices sur Moratiie ou Roanoke River; les guès et chemins au Sud de James River gâtés sous differents pretextes — Mouvements pour occuper l'attention de l'Ennemy. Comme dans l'evenement preparé par le Général la Fayette il seroit resté à la garnison de Portsmouth un moyen d'echapper. le Général fait menacer ce point. le Général Ohara croit devoir enclouer 30 pieces de Canons, et se rejoindre au gros de l'Armée, à peine tout est il reüny que le Comte de Grasse paroit à l'entrée de la Baye de Chesapeak — le Général Waine passe la Rivierre, et se place de maniere à arreter l'ennemy s'il tentoit de se retirer vers la Caroline. l'Amiral Français etoit attendu au Cap Henry par un aide de Camp du Général la Fayette pour lui rendre compte de la situation respective des troupes de terre, et lui demander les mouvements necessaires pour Couper toute retraite aux Ennemis; il mouille au Cap Henry — envoie trois Vaisseaux dans York River, garnit James River de fregattes, et le Mis. de Saint Simon avec 3000 hommes debarque à James Island ou James Town.

Septembre.

La Rivierre ainsy deffendüe, le Général Waine à ordre de la repasser — le Mis. de la Fayette marche sur Willamsburg, reünit dans une bonne position les troupes Combinées au nombre de 7500 hommes environ: il avoit laissé 1500 Miliciens dans le County de Gloucester, et fait hâter quelques troupes venant du Nord — Cette position qui ferme toute retraite à Lord Cornwallis (nos postes avancés à 9 milles d'York) est conservée depuis le 4 jusqu'au 28 Septembre — Lord Cornwallis reconnut la position du Général la Fayette et desespera de la forcer.

Le 6 Septembre le Comte de Grasse laissant les Rivierres gardées, sort avec le reste de sa flotte, poursuit l'Amiral Hood qui s'etoit presenté, le bat et coule à fond le terrible, il prend les Fregattes l'Iris et le Richmond — le 13 il se reünit dans la Baye à l'Escadre de Mr. de Barras partie de Rhode Island avec 800 hommes et l'artillerie française — la Flotte du Comte de Grasse consiste alors en 38 Vaisseaux de ligne.

L'Amiral de Grasse et le Général Saint Simon Commandant les français aux ordres du Général la Fayette, le pressent d'attaquer Lord Cornwallis, et lui offrent un renfort de garnison de Vaisseaux, il prefera agir à Coup sur et d'attendre les troupes venant du Nord. en effet le Général Washington parvint à tromper entierement le Général Clinton sur ses intentions, il s'avançait vers la Virginie avec un detachement Americain, et l'armée du Comte de Rochambeau embarqué sur la tête de Chesapeake, sont portés sur des transports à Willamsburg — le 28 on marcha sur York, et l'armée Combinée en Commence l'Investissement — le 29 reconnoissance de la Place — le 30 l'Ennemy evacüe les postes avancés, et se retire dans les Ouvrages d'York.

Octobre.

Le 1er. Nouvelle reconnoissance — le 3 escarmouche entre la legion de Mr. le Duc de Lauzun, et celle de Tarleton, ou la premiere à l'avantage. Cette legion et 800 hommes des Vaisseaux sous Mr. de Choisy avoient joint la Milice à Gloucester — la nuit du 6 au 7 la tranchée ouverte — celle du 11 au 12 ouverture de la seconde parallelle — la nuit du 14 au 15 les redoutes de la gauche des Ennemis enlevées l'epée à la main, l'une par les grenadiers et Chasseurs français — l'autre par l'Infanterie legere Americaine. la 1ere attaque dirigée par le Baron de Viosmenil Marechal de Camp. la 2eme par le Marquis de la Fayette — le 17 matin Lord Cornwallis demanda à Capituler — le même soir le feu cesse — le 19 l'armée Angloise reduite à 8000 hommes (900 miliciens compris) se rend prisonniere de guerre.

Capitaine

Figure 67. *Carte de la Campagne en Virginie du Major Général Mis. de la Fayette*, Michel Capitaine du Chesnoy, 1781, manuscript, 94 × 148 cm. Map Collection, Yale University Library

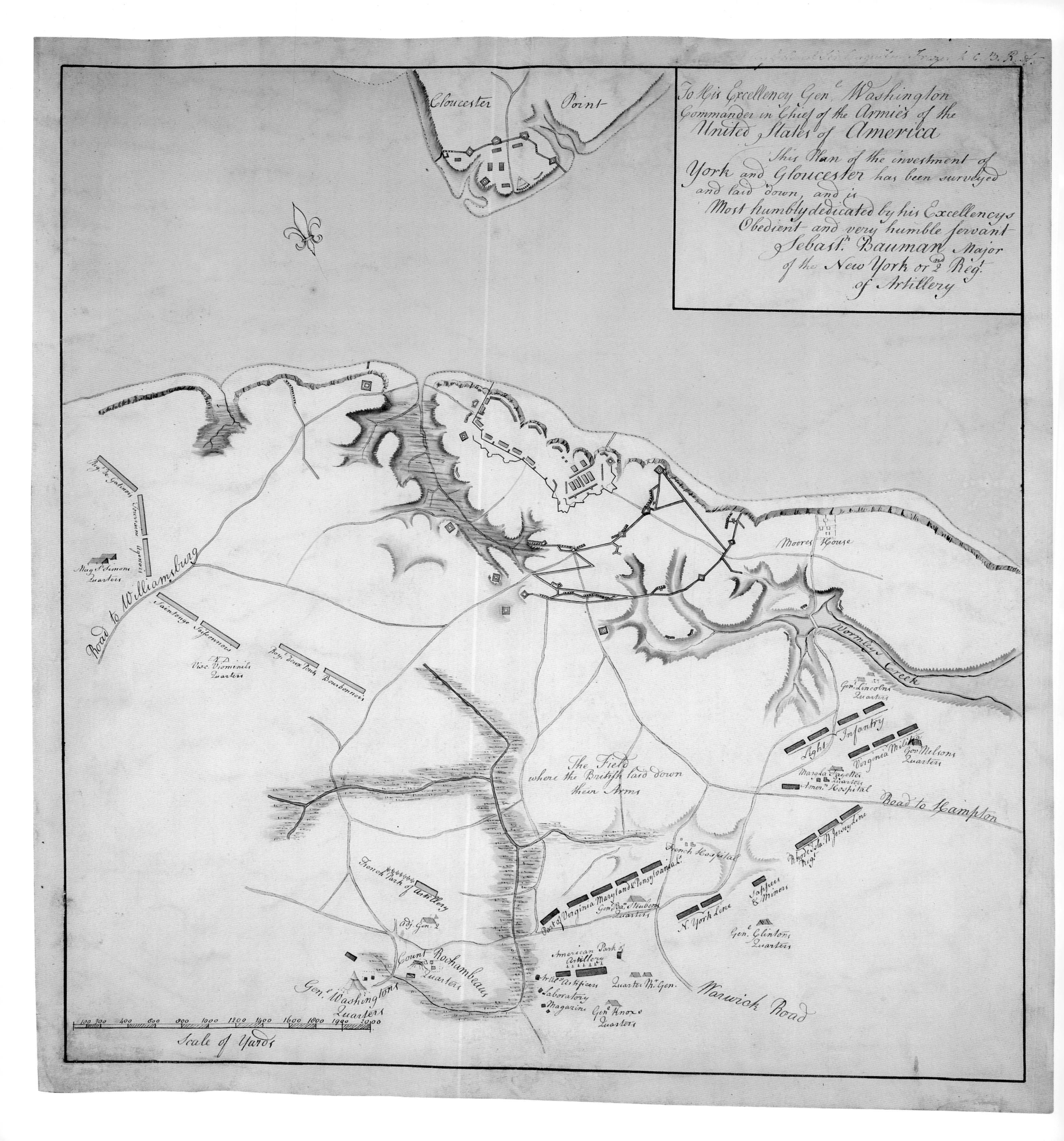

To His Excellency Genl. Washington
Commander in Chief of the Armies of the
United States of America
This Plan of the investment of
York and Gloucester has been surveyed
and laid down, and is
Most humbly dedicated by his Excellencys
Obedient and very humble servant
Sebastn. Bauman Major
of the New York or 2nd Regt.
of Artillery
Gloucester Point
Road to Williamsburg
Touraine
Agenois
Marq. St. Simons Quarters
Saintonge
Soissonnois
Visc. Viomenils Quarters
Roy. Deux Ponts
Bourbonnois
Moores House
Wormleys Creek
Genl. Lincolns Quarters
Light Infantry
Virginia Militia
Gov. Nelsons Quarters
Marq. la Fayettes Quarters
Amern. Hospital
Road to Hampton
The Field where the British laid down their Arms
French Hospital
French Park of Artillery
Part of Virginia Maryland & Pensylvania L.
Genl. Bar. Steubens Quarters
N. York Line
Sappers & Miners
Genl. Clintons Quarters
Count Rochambeaus Quarters
Genl. Washingtons Quarters
American Park of Artillery
Artill. Artificers
Laboratory
Magazine
Quarter Mr. Gen.
Genl. Knoxs Quarters
Warwick Road
Scale of Yards

Society and was a gift from one of Bauman's descendants (figure 70). It is very likely Bauman's own copy of the map. Bauman had announced the publication of the map in the *New-Jersey Journal* of January 20, 1782. There he stated that he had delineated how the British "posts were besieged in form by the allied army of America and France; the British lines of defense, and the American and French lines of approach; with part of York River, and the British ships, as they then appeared sunk in it before York-Town; and the whole encampment in its vicinity."

Washington's importance as commander in chief is accentuated by the enormous size of his tent in the lower left quadrant of the map. This tent survived the battle and was sometimes erected at patriotic events, such as during Lafayette's visit to America in 1824. It can still be seen at the Museum of the American Revolution in Philadelphia. Rochambeau's tent is also indicated as well as the quarters of General Lafayette near Wormley Creek. "The Field where the British laid down their Arms" is at the center of the map.

Lafayette's personal mapmaker, Michel Capitaine du Chesnoy, executed the large and beautifully colored map of the Yorktown campaign purchased for the Yale library. Three manuscripts of the map of the Virginia campaign are known to exist, all of them in libraries in the United States. In addition to Lafayette's copy at Yale, there are also examples at Colonial Williamsburg and the Library of Congress.

On the map, Capitaine documented the sequence of events leading up to the final battle at Yorktown—troop disposition and movement, terrain, location of camps and batteries. A précis of the campaign of 1781 is written on the left side of the map where Capitaine details the important role Lafayette played at Yorktown. The final paragraph reads, "On the first [of October] an attack directed by Baron de Viominil, Camp Marshal, the second [of October] by the Marquis de La Fayette."

The morning of October 17 Lord Cornwallis (figure 69) asked to surrender and the same evening the firing stopped. On the nineteenth the English army, reduced to about eight thousand men including nine hundred militia, became prisoners of war. As commander of American troops in Virginia in the spring and summer of 1781, "Lafayette did more than any other allied commander to prepare the way for the British capitulation at Yorktown. The greatest American victory was Lafayette's most shining personal triumph," wrote Lafayette's biographer. Lafayette's dedication to the cause of liberty, like that of Washington, whom he so admired, made him one of the greatest heroes in American history. The deeds of a hero must be recorded and Lafayette had the good sense to bring along his own mapmaker for the purpose. Capitaine served his commander well by creating remarkable graphic documents of Lafayette's extraordinary accomplishments. His crowning achievement is this map of Virginia.

In 1824, Lafayette made a second triumphant trip to America where he was lionized as he traveled. He was back in France in 1825 and an American historian visited him at La Grange, his ancestral home. Jared Sparks, the editor of the *North American Review* and a future president of Harvard College, was in Europe on a research trip, collecting and copying documents relating to the American Revolution. One of his missions was to see the memorabilia in Lafayette's possession. The general gallantly received the historian and laid his cherished maps before him. They showed "all the actions in which he was engaged

Figure 69. *Portrait of Charles, 1st Marquis Cornwallis*, Daniel Gardner, 1782, pencil, pastel, and bodycolor, oval, 90 × 66 cm. © Christie's Images/Bridgeman Images

Figure 68. *To His Excellency Genl. Washington, Commander in Chief of the Armies of the United States of America, this plan of the investment of York and Gloucester has been surveyed and laid down*, Sebastian Bauman, 1781, manuscript, 49 × 45 cm. © The British Library Board [ADD. 57715 (Part 13)]

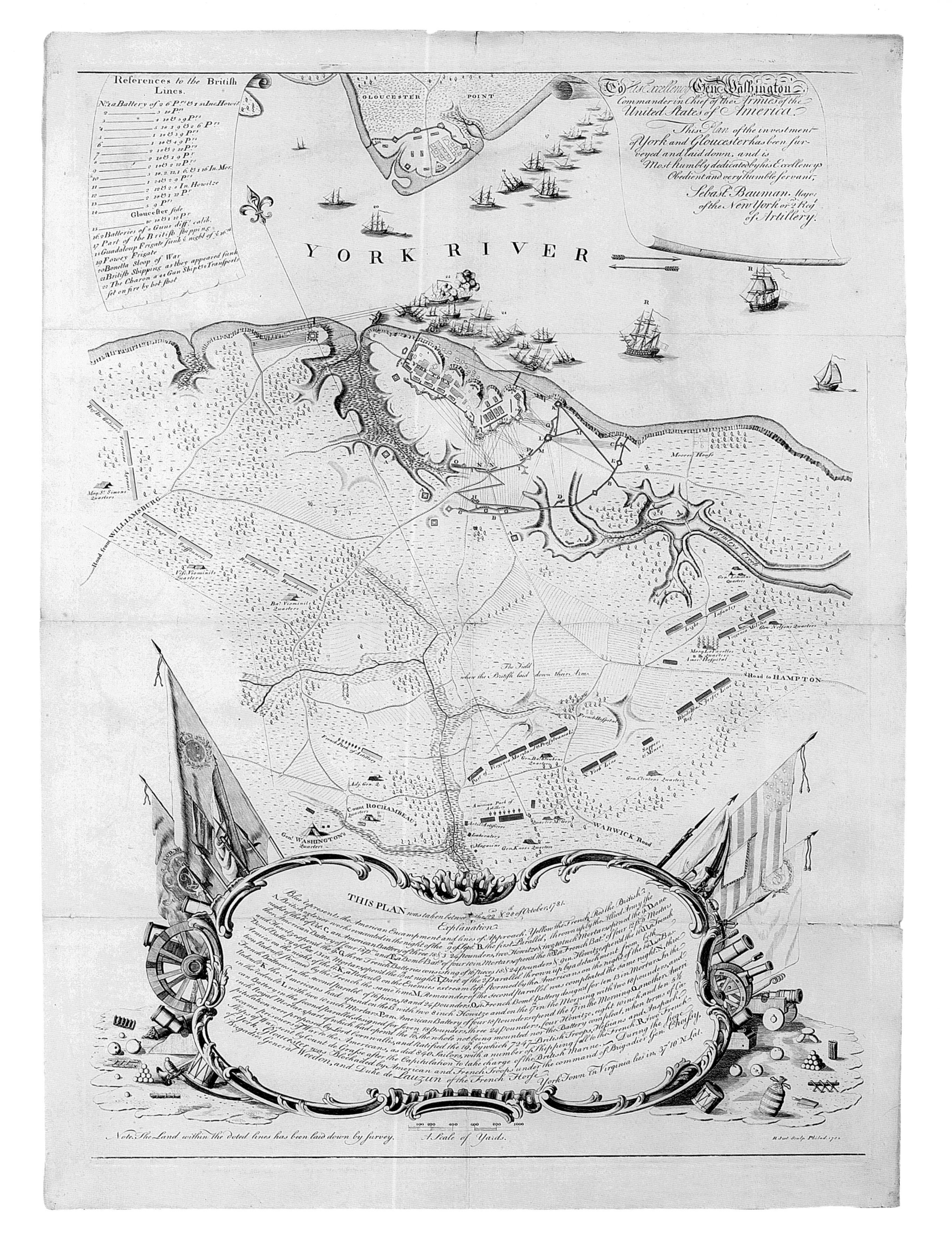
References to the British Lines.
Gloucester side
16 2 Batteries of 8 Guns diff.t calib.
17 Part of the British shipping
18 Guadaloup Frigate sunk y.e night of y.e 16.th
19 Fowey Frigate
20 Bonetta Sloop of War
21 British Shipping as they appeared sunk
22 The Charon a 44 Gun Ship & 2 Transports set on fire by hot shot
GLOUCESTER POINT
To His Excellency Gen.l Washington Commander in Chief of the Armies of the United States of America.
This Plan of the investment of York and Gloucester has been surveyed and laid down, and is Most humbly dedicated by his Excellency's Obedient and very humble servant,
Sebast.n Bauman. Major of the New York or 2.d Reg.t of Artillery.
YORK RIVER
Road from WILLIAMSBURG
Wormley's Creek
Moores House
Road to HAMPTON
WARWICK Road
Count ROCHAMBEAU
Gen.l WASHINGTON'S Quarters
THIS PLAN was taken between the 22 & 28 of October 1781.
Explanation
York Town in Virginia lies in 37° 10' N. Lat.
Note. The Land within the doted lines has been laid down by survey.
A Scale of Yards.

in America," and Sparks was so impressed that he requested copies made from the originals. Under the watchful eye of Lafayette himself, some or all of the maps that had been drafted and colored by Capitaine were copied with meticulous care. Six of these copies are in the Cornell University library in Ithaca, New York. "The General has . . . a map of the Virginia campaigne, taken at the time," wrote Sparks, who reported that he had this map copied, although it is not among the maps at Cornell.

The map that Vietor bought for Yale was originally mounted on sticks of blue and gold and kept in a specially made blue cardboard box. Lafayette liked to remove it from the box with great ceremony when he showed it to the likes of Sparks. There were moments of concern as the map made its way from Europe to New Haven. It was mailed from Switzerland, but two months went by and it had not been delivered. "I was very worried," wrote Vietor, "lest by some method it had gone on the ANDREA DORIA." That was the Italian ocean liner that sank near Nantucket on July 25, 1956, after colliding with the MS *Stockholm*. While urgent requests for payment were being wired from France, a mail clerk from the Yale postal station calmly appeared at the map library carrying a package that had been languishing at the post office for weeks because fifteen cents postage was due. No notice of the delinquent funds had been sent to the library and without payment the package could not be delivered. Rather than return it to Switzerland, the clerk had gone directly to the map department to collect the change himself. "A short time thereafter a fire broke out near the post office and the entire place was water damaged!" wrote Vietor. "By such a slim margin did we get the map safely at Yale." The national treasure arrived in time to serve as the centerpiece of an exhibition at Yale to commemorate the two hundredth anniversary of Lafayette's birth.

Peace

When Richard Oswald, representing Great Britain, and the Americans John Adams, Benjamin Franklin, John Jay, and Henry Laurens signed the preliminary articles of peace on November 30, 1782, the United States was well on its way to gaining clear title to vast lands in North America. Lord Shelburne, the secretary of colonial affairs, had selected Oswald to travel to Paris for the negotiations in part because he had lived in America for many years and was well known for his many friendships with prominent businessmen throughout the colonies. He had built one of the largest slave-trading empires and his business acumen had earned him the respect of many, including Franklin, who admired his air of "great simplicity and honesty."

On December 6, 1782, the peace commissioners laid a large map on a table and Oswald took up his pen and began to draw the new boundaries of the United States. In bright red ink, he laid down the borders with France in Louisiana and Canada and with Spain in Florida. "I now can assure you," Franklin wrote to Thomas Jefferson, "that I am perfectly clear in Remembrance that the map we used in tracing the Boundary [on

Figure 70. *To His Excellency Genl. Washington, Commander in Chief of the Armies of the United States of America, this plan of the investment of York and Gloucester has been surveyed and laid down*, Sebastian Bauman, Philadelphia, 1782, copperplate engraving, 66 × 48 cm. Collection of the New-York Historical Society

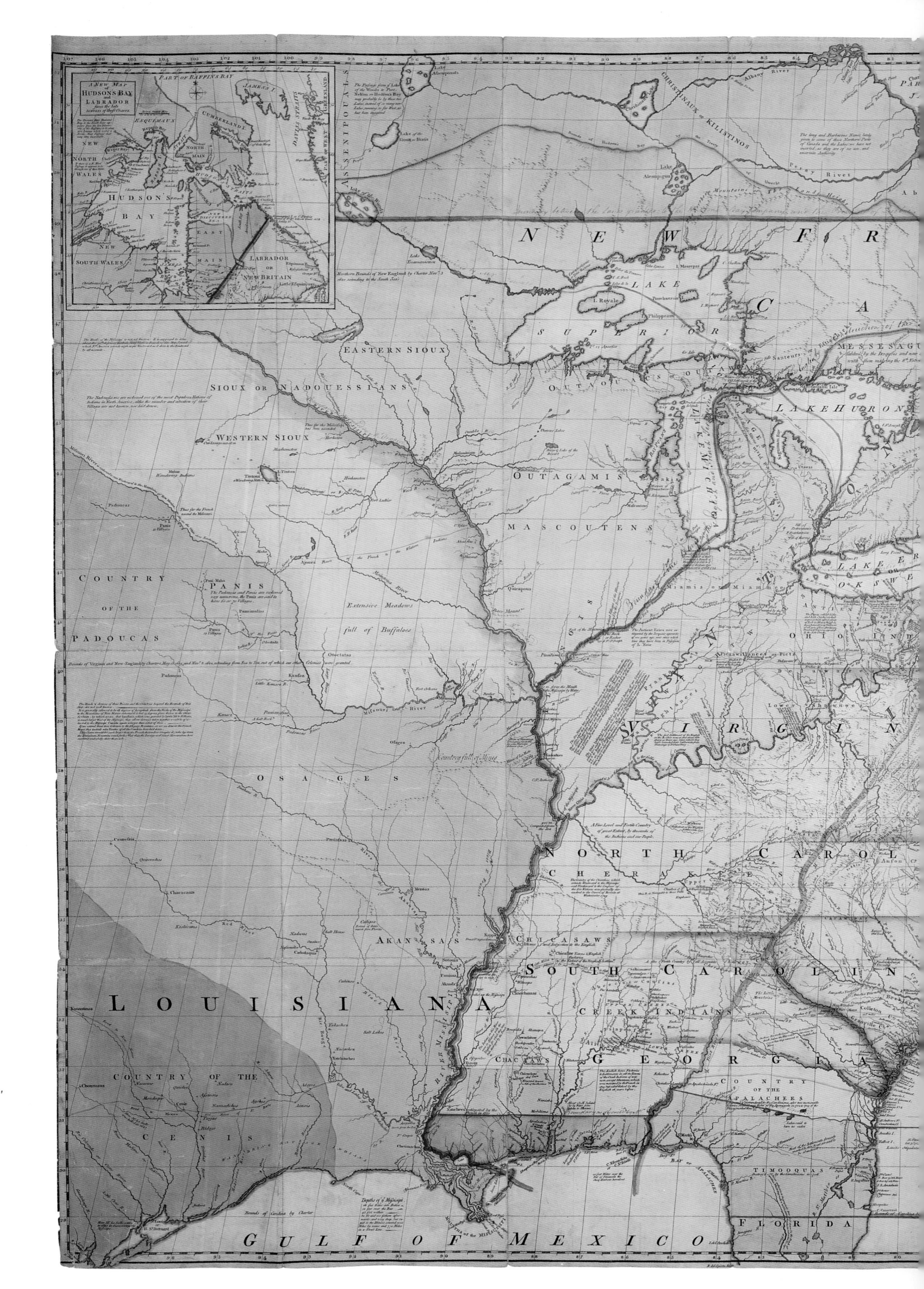

Figure 71. *A Map of the British Colonies in North America*, John Mitchell, London, 1755/1775, copperplate engraving with manuscript additions by Richard Oswald, ca. 1782, 197 × 140 cm. © The British Library Board (Maps K. Top 118.49b)

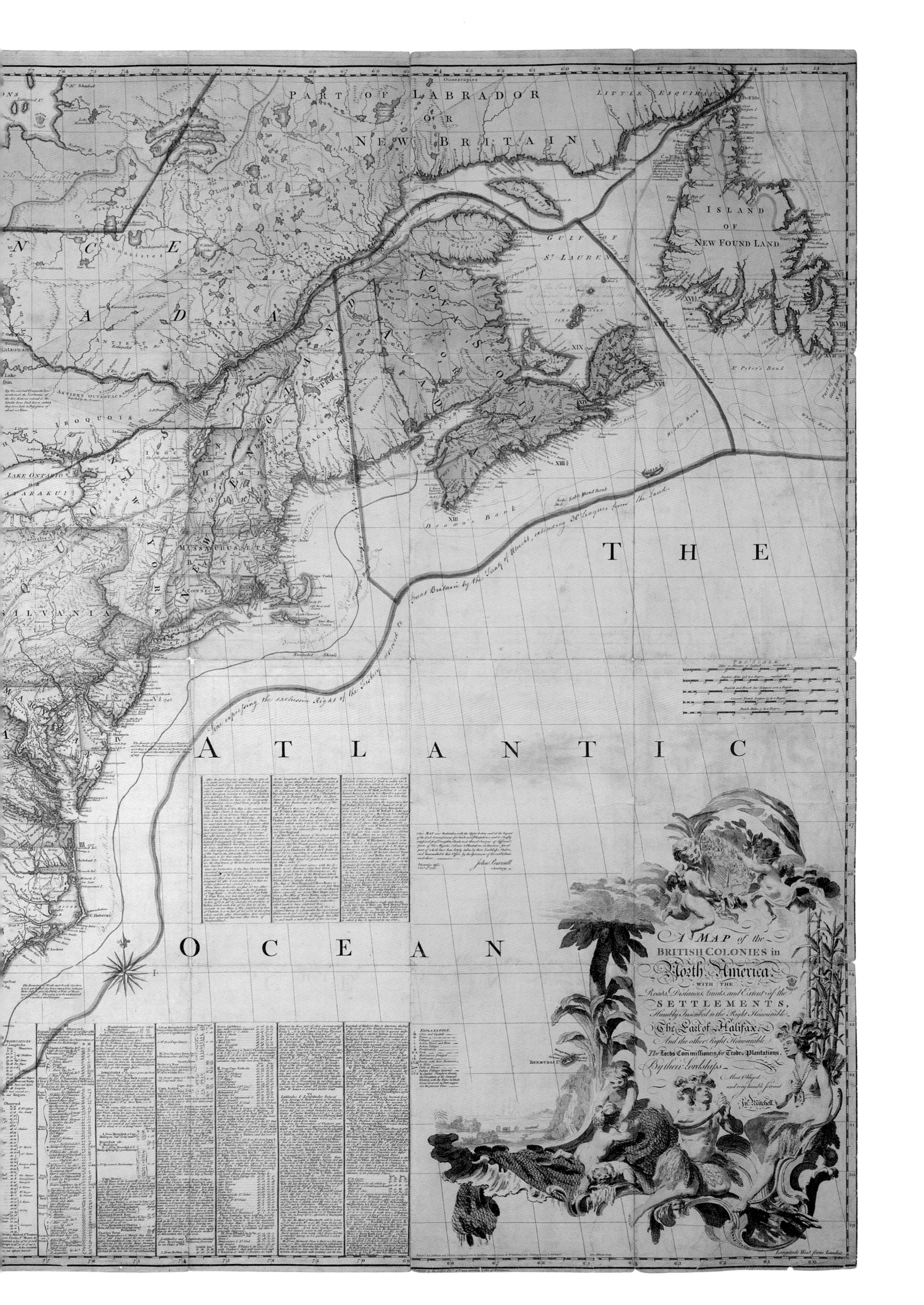

PART OF LABRADOR
OR
NEW BRITAIN
LITTLE ESQUIMAUX
ISLAND
OF
NEW FOUND LAND
GULF OF
ST LAURENCE
NOVA SCOTIA
IROQUOIS
LAKE ONTARIO
OR
CATARAKUI
MASSACHUSETTS
BAY
NEW YORK
Brown's Bank
Line expressing the exclusive Right of the Fishery secured to Great Britain by the Treaty of Utrecht, extending 30 Leagues from the Land.
THE
ATLANTIC
OCEAN
BERMUDAS
A MAP of the
BRITISH COLONIES in
North America
WITH THE
Roads, Distances, Limits, and Extent of the
SETTLEMENTS,
Humbly Inscribed to the Right Honourable
The Earl of Halifax,
And the other Right Honourable
The Lords Commissioners for Trade & Plantations,
By their Lordships
Most Obliged
and very humble Servant
Jno. Mitchell

December 6, 1782] was brought to the Treaty by the Commissioners from England, and it was the same that was published by Mitchell 20 years before."

John Mitchell's map had actually been published twenty-seven years earlier, at the time when England and France were disputing the claims in North America that caused the French and Indian War. Mitchell's map was nevertheless the obvious choice for the commissioners. His *Map of the British Colonies in North America* (1755) was far and away the most significant of America for the second half of the eighteenth century, and there was no newer map of North America that was more up-to-date. In fact, almost all of the other maps of the colonies published between 1755 and 1782 were simply derivatives of the Mitchell map.

The example from the British Library (figure 71) illustrated on pages 138–39 is one of the maps annotated by Oswald. Called the "Red-lined map," it has a boundary added in manuscript that travels up the St. Lawrence River, through the Great Lakes, and down the Mississippi River. A second red-lined map is at the New-York Historical Society and may have been annotated at a different time because the boundaries are not exactly the same and it is a different state of the map. For example, on the New-York Historical Society's map, the red line crosses Lake Superior dividing the fictitious Isle Phillippeaux equally between the French and the Americans. On the British Library's copy the entire island went to the Americans.

Within the lines on each map was the vast trans-Appalachian West, which was incorporated as part of the new United States. The British legation attending the map-annotating meeting had been prepared to concede more than the territory specified in the preliminary treaty of peace. Nevertheless, many officials in England thought the terms were too generous. The American commissioners could not have been more pleased. The spoils of war had resulted in an enormous new country.

John Mitchell was a native American, born in Virginia, but he was educated at the University of Edinburgh, Scotland. He became one of the most accomplished men of his generation: "He has a place in the early history of American botany, zoology, physiology, medicine, cartography, climatology, and agriculture, to say nothing of politics. He was for a time perhaps the ablest scientific investigator in North America."

At the conclusion of the War of the Spanish Succession (Queen Anne's War), the territorial delimitations for North America laid out in the 1713 Treaty of Utrecht were vague, permitting the British and French great leeway in interpreting their entitlements. At the same time that the British were claiming all the territory from the Atlantic to the Mississippi between the Gulf of Mexico and the Great Lakes, France maintained that her colonies occupied the area west of the Appalachian Mountains, including the Ohio Valley and Louisiana. The French backed up their claims by constructing menacing forts along the Ohio and Mississippi Rivers in order to thwart westward expansion by the British.

This state of affairs dismayed Mitchell who lamented the fact that the only large, detailed map of North America then available was a 1733 work by Henry Popple that did not address itself to the confusion over territorial claims. As the French continued to encroach upon the British colonies, the chauvinistic Mitchell decided to publish an impressive map to demonstrate the mounting threat to the British colonies in America. He may have had the map in his mind as early as 1746 when, as an ailing scientist, he set sail for

England to regain his health. At that time, the British were battling the French in the War of the Austrian Succession, and a French privateer captured Mitchell's ship. His hostility toward the French was not diminished after suffering as one of their prisoners. Soon after undergoing this harrowing experience, he and his wife arrived safely in London where he turned his attention to the cartography of America.

For five years, Dr. Mitchell collected every scrap of geographical information he could lay his hands on in order to create a comprehensive and up-to-date map that would have no peer. He tracked down every geographer and historian and told one correspondent that "there are none I believe but what I have consulted." Other resources were the printed maps of the 1740s and '50s, but he also had the rare privilege to examine official documents. Through the good offices of his friend George Montagu Dunk, Earl of Halifax, he was given entree to the repository of manuscript maps and geographical materials on file in the archives of the Board of Trade in London. Mitchell's original purpose was to show the division of North America between the British and the French, but the resulting map, dedicated to Dunk, is so detailed and accurate that it has been used to resolve border disputes as recently as the twentieth century.

At the western extremities of the map, Mitchell cited charters dated May 23, 1609, and November 3, 1620, which stated that the western boundaries of Virginia and New England stretched "from Sea to Sea, out of which our other colonies were granted." In these few words, Mitchell cavalierly claimed for the British all the vast, unexplored lands in North America to the Pacific Ocean. Although the map itself does not incorporate land in the far west, a part of the Louisiana territory is delineated, extending into present-day Oklahoma, Kansas, and Nebraska. At that time, the French and the Spanish possessed the lands west of the Mississippi, but Mitchell suggested that this unexplored geography would one day be part of a unified country.

Like most of the maps of the period, the western boundaries of such British colonies as Georgia, North Carolina, and Virginia extend west at least to the Mississippi River. On some examples, the coloring of the colonies extends west of the Mississippi, presumably all the way to the Pacific. The map is a compendium of information about America with extensive notes, topographical features, locations of Indian tribes, dates of various settlements, and principal roads.

John Mitchell's only published map quickly became the standard of North America, and twenty-one distinct editions were eventually printed with updated geographical information and in different languages. The map itself has been put into service many times to settle border disputes, but it never had such prominent use than during the negotiations for peace at the conclusion of the American Revolutionary War. The copy illustrated here was given to King George III and must have been a stinging reminder to him of his tremendous losses in North America. It is now one of the treasures of the British Library. John Jay took possession of the second red-lined map, and it remained in his family until it was donated to the New-York Historical Society by its eighth president, Peter Jay. The Mitchell has long been considered the most important map in American history.

Acknowledgments

Revolution is the result of a collaboration between a dealer and a collector of rare maps. Richard Brown's collection of Revolutionary War maps can be viewed on the website at the Norman Leventhal Map Center at the Boston Public Library (http://maps.bpl.org/highlights/ar/richard-h-brown-revolutionary-war-map-collection). It was the basis and inspiration for this book. At the Center, we thank Jan Spitz, the executive director; Ron Grim, curator; Stephanie Cyr, assistant curator; and Tom Blake, digital project manager.

On October 4, 2012, we visited Alnwick Castle in Northumberland, England, to see the Revolutionary War maps that had belonged to Lord Hugh Percy, an active combatant in the early battles of the war. *Revolution* took form in our minds after visiting Alnwick. Here were many of the most intriguing maps of the war and very few people knew of their existence; many had never been reproduced before. We thank Christopher Hunwick, the Alnwick archivist for his help and Eve Reverchon for providing digital images of the maps.

On that same trip to England, we spent several days at the British Library examining the works in the King George III Topographical Collection. We are grateful to Peter Barber, head of Map Collections, and Tom Harper, curator of Antiquarian Maps. Their knowledge and expertise were indispensable as we looked at every map that related to the French and Indian War and the Revolution. We thank them for their guidance and patience. We also found important manuscript maps in the Manuscript Division of the British Library.

We inspected nearly every manuscript map relating to these wars at the Library of Congress. Patricia Molen van Ee, the co-compiler of *Maps and Charts of North America and the West Indies, 1750 to 1789* spent long hours with us in the vault of the Geography and Map library. Ralph Ehrenberg, Robert Morris, and Edward Redmond constantly provided professional advice. Diane Schug-O'Neill provided assistance with the many fine reproductions. We thank them as well as Susan Siegel, director of development at that great national library.

At the American Antiquarian Society we thank Ellen Dunlap, Matthew Shakespeare, and Lauren Hewes, and we thank Abraham Parrish of the Yale Map Library. At the New-York Historical Society, Mariam Touba replied to our numerous questions about materials in that venerable institution. Mary Pedley and Brian Dunnigan spent two days with us at the Clements Library, University of Michigan, as we went through the maps in the collections of generals Thomas Gage and Henry Clinton. Ed Dahl, former curator of maps at the National Archives of Canada was also very helpful.

We have enjoyed many months of working with our editor, Jim Mairs of W. W. Norton. We thank him for his perceptive suggestions and most of all his indulgence with us. Robert Karrow read the manuscript and made many useful suggestions. Our gratitude goes to Don Kennison for wise copyediting. We thank Laura Lindgren, who masterfully designed and typeset the book. Julia Druskin of W. W. Norton expertly managed the book's production.

Our agent, Asher Jason, gave us good advice and guidance from start to finish.

On a personal note, we wish to thank Phyllis Posnick, Mary Jo Otsea, and Zach Johnk who supported this project from the beginning.

Bibliography

Alden, John R. *General Gage in America.* Baton Rouge: Louisiana State University Press, 1948.

———. *A History of the American Revolution.* New York: Knopf, 1969.

Allan, D. G. C. "The Laudable Association of Antigallicans," *RSA Journal* 137 (London, 1989).

Allen, Gardner Weld. *A Naval History of the American Revolution,* 2 volumes. Boston: Houghton Mifflin Company, 1913.

Anderson, Fred. *Crucible of War.* New York: Vintage Books, 2001.

Argenson, René-Louis de Voyer. *Journal and Memoirs of the Marquis d'Argenson,* Katharine Prescott Wormeley, trans. Cambridge, MA: University Press, 1901.

Arnold, Howard Payson. *The Washington Medal; In Commemoration of the Evacuation of Boston 17 March 1776.* Boston: Hardy, Pratt & Company, 1976.

Bearss, Edwin C. *The Battle of Sullivan's Island and the Capture of Fort Moultrie.* Washington, D.C.: National Park Service, 1968.

Blake, Henry Taylor. *The Battle of Lake George.* N.p.: Nabu Press, 2014.

Boatner, Mark M. *Encyclopedia of the American Revolution.* New York: David McKay Company, Inc., 1974.

Boscawen, Hugh. *The Capture of Louisbourg, 1758.* Norman: University of Oklahoma Press, 2011.

British Library. *American War of Independence.* London: British Museum Publications Ltd., 1975.

Brodhead, John Romeyn. *Documents Relative to the Colonial History of the State of New York.* Albany: Weed, Parsons and Company, 1856.

Brumwell, Stephen. *Paths of Glory.* London: Hambledon Continuum, 2006.

Brun, Christain. *Guide to the Manuscript Maps in the William L. Clements Library.* Ann Arbor: University of Michigan, 1959.

Burgoyne, Bruce E., trans. *Defeat, Disaster, and Dedication: The Diaries of the Hessian Officers Jakob Piel and Andreas Wiederholdt.* Westminster, MD: Heritage Books, Inc., 2008.

Clinton, Sir Henry. *The American Rebellion, Sir Henry Clinton's Narrative of his Campaigns, 1775–1782.* William B. Willcox, ed. New Haven: Yale University Press, 1954.

Cohen, Paul E., and Robert T. Augustyn. *Manhattan in Maps, 1527–1995.* New York: Rizzoli, 1997.

Cohen, Paul E. "Michel Capitaine du Chesnoy, the Marquis de Lafayette's cartographer," *The Magazine Antiques,* CLIII (January 1998), 170–77.

Cooper, James Fenimore. *The Last of the Mohicans.* Philadelphia: H. C. Carey & I. Lea, 1826.

Creasy, Edward. *Fifteen Decisive Battles of the World.* London: Richard Bentley, 1852.

Crocker, Thomas E. *Braddock's March.* Yardley, PA: Westholme, 2009.

Cumming, William P. *British Maps of Colonial America.* Chicago: University of Chicago Press, 1974.

———. *The Fate of a Nation.* London: Phaidon Press, 1975.

———. "The Treasures of Alnwick Castle," *American Heritage* XX (August 1969): 23–26 and 99–101.

Desjardin, Thomas A. *Through a Howling Wilderness.* New York: St. Martin's Press, 2006.

Dickens, Charles. *American Notes for General Circulation.* New York: Harper & Brothers, 1842.

Downey, Fairfax. *Louisbourg: Key to a Continent.* Englewood Cliffs, NJ: Prentice-Hall, Inc., 1955.

Dupuy, R. Ernest. *The Compact History of the Revolutionary War.* New York: Hawthorn Books, 1963.

Dwyer, William M. *This Day Is Ours!: An Inside View of the Battles of Trenton and Princeton.* New York: The Viking Press, 1983.

Fowler, William M. Jr. *Empires at War.* New York: D&M Publishers, Inc., 2005.

Franklin, Benjamin. *The Autobiography of Benjamin Franklin.* New York: Tribeca Books, 2013.

Gage, Thomas. *The Correspondence of General Thomas Gage with the Secretaries of State, and with the War Office and the Treasury 1763–1775.* New Haven: Yale University Press, 1933.

Garden, Alexander. *Anecdotes of the Revolutionary War in America.* Charleston: A. E. Miller, 1822.

Gibbes, Robert Wilson. *Documentary History of the American Revolution: 1776–1782.* New York: D. Appleton & Co., 1857.

Gipson, Lawrence Henry. *The British Empire Before the American Revolution,* volumes 6–11. New York: Alfred A. Knopf, 1961–65.

Gordon, John D. *South Carolina and the American Revolution.* Columbia: University of South Carolina Press, 2003.

Guthorn, Peter J. *American Maps and Map Makers of the Revolution.* Monmouth Beach, NJ: Philip Freneau Press, 1966.

———. *British Maps of the American Revolution.* Monmouth Beach, NJ: Philip Freneau Press, 1972.

Hadden, James Murray. *Hadden's Journal and Orderly Books: A Journal Kept in Canada and upon Burgoyne's Campaign in 1776 and 1777.* Albany: J. Munsell's Sons, 1884.

Harley, J. B. *Mapping the American Revolutionary War.* Chicago: University of Chicago Press, 1978.

Hibbert, Christopher. *Wolfe at Quebec*. New York: World Pub. Co., 1959.

Hornberger, Theodore. "The Scientific Ideas of John Mitchell," *Huntington Library Quarterly* 10 (1946–47): 277–96.

Hornsby, Stephen J. *Surveyors of Empire*. Ithaca, NY: McGill-Queen's University Press, 2011.

Horry, Peter, and M. L. Weems. *The Life of General Francis Marion*. Philadelphia: J. B. Lippincott & Co., 1884.

Hulbert, A. B. *Historic Highways of America: Braddock's Road*. Cleveland: The Arthur H. Clark Company, 1903.

Kalm, Peter. *The Travels of Peter Kalm, Finnish-Swedish Naturalist, Through Colonial North America, 1748–1751*. Fleischmanns, NY: Purple Mountain Press, 2007.

Kooperman, Paul. *Braddock at the Monongahela*. Pittsburgh, PA: University of Pittsburgh Press, 1992.

Lacey, James. *Moments of Battle*. New York: Random House, 2013.

Lancaster, Bruce. *The American Heritage Book of the Revolution*. New York: American Heritage Publishing Co., Inc., 1958.

Lawrence, Alexander A. *Storm over Savannah*. Athens: University of Georgia Press, 1951.

Library of Congress. *Maps and Charts of North America and the West Indies, 1750–1789, compiled by John R. Sellers and Patricia Molen van Ee*. Washington, D.C.: Library of Congress, 1981.

Lie of the Land: The Secret Life of Maps. April Carlucci and Peter Barber, eds. London: British Library, 2001.

Mackellar, Patrick. MacKellar's *Journal* is published in the appendix of Arthur George Doughty, *The Siege of Quebec and the Battle of the Plains of Abraham*. Quebec: Dussault & Prouix, 1901, pp. 33–58. Mackellar's *Journal* is mistakenly attributed to Major James Moncrief in the appendix.

Mahan, A. T. *The Major Operations of the Navies in the War of American Independence*. London: Sampson, Low, Marston & Company, 1913.

Marshall, Douglas W., and Howard Peckham. *Campaigns of the American Revolution*. Ann Arbor: The University of Michigan Press, 1976.

McBurney, Christian, M. *Rhode Island Campaign*. Yardley, PA: Westholme, 2011.

McCrady, Edward. *The History of South Carolina in the Revolution*. New York: The Macmillan Company, 1901.

McNairn, Alan. *Behold the Hero: General Wolfe and the Arts in the Eighteenth Century*. Montreal: McGill-Queen's University Press, 1997.

Montresor, John. *The Montresor Journals*, G. D. Scull, ed. *Collections of the New-York Historical Society for the Year 1881* (New York, 1882).

Morgan, Edmund S., and Helen M. Morgan. *The Stamp Act Crisis: Prologue to Revolution*. Chapel Hill: The University of North Carolina Press, 1953.

Morrill, Dan L. *Southern Campaigns of the American Revolution*. Baltimore: Nautical & Aviation Pub. Co., 1993.

Moultrie, William. *Memoirs of the American Revolution*. New York: David Longworth, 1802.

Nebenzahl, Kenneth. *Atlas of the American Revolution*. Chicago: Rand McNally, 1974.

———. *A Bibliography of Printed Plans of the American Revolution*. Chicago: The University of Chicago Press, 1975.

Nichols, Francis. "Diary of Lieutenant Francis Nichols," *The Pennsylvania Magazine of History and Biography* 20 (1896).

O'Conner, Antoine François Térence, and Charles Henri, comte d'Estaing. *Journal du Siege de Savannah Septembre, et Octobre 1779* (1779). Manuscript, John Carter Brown Library (1-SIZE Codex Fr 11).

Palmer, Peter Sailly. *Battle of Valcour on Lake Champlain, October 11th, 1776*. Rouses Point, NY: Lake Shore Press, 1876.

Pargellis, Stanley. *Military Affairs in North America 1748–1765*. New York: D. Appleton-Century Company, 1936.

Parkman, Francis. *France and England in North America*. New York: The Library of America, 1983.

Pedley, Mary. "A New and Accurate Map of the English Empire in North America by a Society of Anti-Gallicans (London, 1755)." *Mappae Antiquae: Liber Amircorum Günter Schilder*. Leiden: Hes & De Graaf, 2007.

Piecuch, Jim. *The Battle of Camden*. Charleston, SC: History Press, 2006.

Pritchard, Margaret Beck, and Henry Taliaferro. *Degrees of Latitude: Mapping Colonial America*. New York: Henry N. Abrams, 2002.

Raddall, Thomas Head. *Halifax, Warden of the North*. Garden City, NY: Doubleday, 1965.

Ramsay, David. *The History of the Revolution of South Carolina*. Trenton: Isaac Collins, 1785.

Rice, Howard C., Jr., and Anne S. K. Brown, trans. and eds. *The American Campaigns of Rochambeau's Army 1780, 1781, 1782, 1783*. Princeton: Princeton University Press, 1972.

Roberts, Kenneth. *March to Quebec, Journals of the Members of Arnold's Expedition*. New York: Doubleday, Doran & Company, 1938.

Russell, David Lee. *The American Revolution in the Southern Colonies*. Jefferson, NC: McFarland & Co., 2000.

———. *Victory on Sullivan's Island*. Haverford, PA: Infinity Publishing Company, 2002.

Ruville, Albert von. *William Pitt, Earl of Chatham*. London: W. Heinemann, 1907.

Sandby, Paul. *Picturing Britain*. London: Royal Academy of Arts, 2009.

Sargent, Winthrop, ed. *The History of an Expedition against Fort Du Quesne, in 1755: Under Major-General Edward Braddock*.

Philadelphia: For the Historical Society of Pennsylvania, 1856. Includes Robert Orme's diary.

Schecter, Barnet. *The Battle for New York*. New York: Walker, 2002.

Seelye, Elizabeth Eggleston. *Lake George in History*. Lake George, NY: Elwyn Seelye, 1896–97.

Senter, Isaac. *The Journal of Isaac Senter*. Philadelphia: Historical Society of Pennsylvania, 1846.

Spendlove, F. St. George. *The Face of Early Canada*. Toronto: Ryerson Press, 1958.

Stacey, C. P. *Quebec, 1759: The Siege and the Battle*. Quebec: Robin Brass Studio, 2002.

Stedman, Charles. *History of the Origin, Progress, and Termination of the American War*. London: Printed for the Author, 1794.

Symonds, Craig L. *A Battlefield Atlas of the American Revolution*. Annapolis, MD: Nautical & Aviation Pub. Co. of America, 1968.

Thackeray, Francis. *A History of Right Honourable William Pitt*. London: Printed for C. and J. Rivington, 1827.

Trumbull, John. *The Autobiography of Colonel John Trumbull*, Theodore Sizer, ed. New Haven: Yale University Press, 1953.

United States Continental Congress. *Journals of Congress Containing the Proceedings from January 1, 1776 to January 1, 1777*. Philadelphia: David C. Claypole, 1778.

Wahll, Andrew J. *Braddock Road Chronicles*. Bowie, MD: Heritage Books, 1999.

Williams, Richard. *Discord and Civil Wars (from the original journal)*. Buffalo, NY: Easy Hill Press for the Salisbury Club of Buffalo, 1954.

Willson, Beckles. *The Life and Letters of James Wolfe*. London: W. Heinemann, 1909.

Wilson, David K. *The Southern Strategy: Britain's Conquest of South Carolina and Georgia*. Columbia: University of South Carolina Press, 2005.

Wood, William Charles Henry. *The Great Fortress: A Chronicle of Louisbourg, 1720–1760*. Toronto: Brook & Company, 1915.

Index

Page reference in *italics* refer to illustrations.